6th International Congress on Sustainability Science and Engineering (ICOSSE'17)

Barcelona, Spain
2 - 4 October 2017

ISBN: 978-1-5108-7283-7

Printed from e-media with permission by:

Curran Associates, Inc.
57 Morehouse Lane
Red Hook, NY 12571

Some format issues inherent in the e-media version may also appear in this print version.

Printed by Curran Associates, Inc. (2018)

For permission requests, please contact AIChE
at the address below.

AIChE
120 Wall Street, FL 23
New York, NY 10005-4020

Phone: (800) 242-4363
Fax: (203) 775-5177

www.aiche.org

Additional copies of this publication are available from:

Curran Associates, Inc.
57 Morehouse Lane
Red Hook, NY 12571 USA
Phone: 845-758-0400
Fax: 845-758-2633
Email: curran@proceedings.com
Web: www.proceedings.com

TABLE OF CONTENTS

SESSION 1: THE TRANSITION TO SUSTAINABLE LIVING 1

SESSION 2: THE TRANSITION TO SUSTAINABLE LIVING 2

SESSION 3: SUSTAINABLE TECHNOLOGIES 1

SESSION 4: SUSTAINABLE TECHNOLOGIES 2

ICOSSE PLENARY LECTURE – RENATE CHRIST

SESSION 5: BRINGING SUSTAINABILITY TO THE MAINSTREAM 1

SESSION 6: BRINGING SUSTAINABILITY TO THE MAINSTREAM 2

ICOSSE PLENARY LECTURE – JACQUELINE CRAMER

SESSION 7: BRINGING SUSTAINABILITY TO THE MAINSTREAM 3

Delivering Resource Projects Sustainably in Emerging Economies:

The success story of an innovative business intervention model

Lauren Basson[1], Mike Mulcahy[1], Jim Petrie[2]

1. Green Cape www.green-cape.co.za
2. University of Sydney / University of Cape Town

WCCE/ICOSSE Barcelona 2017

Resource management and infrastructure investments often fail:

NOT because of:

- Lack of understanding of sustainable development
- Lack of clever ideas or smart technologies
- Lack of requisite skills

GreenCape | 2

Resource management and infrastructure investments often fail:

BUT because of:

- Misalignment of government policy
- Efficacy of delivery agencies
- Regulatory, commercial and financial risks
- Lack of effective stakeholder communication

GreenCape was enabled by three macro-trends dovetailing

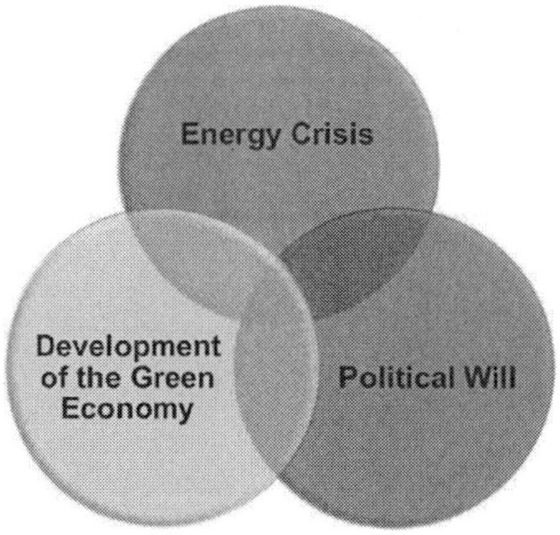

Political will

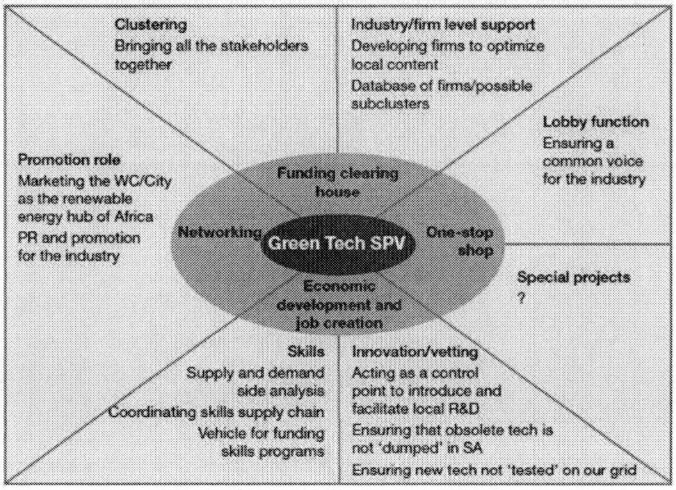

Western Cape Government

- Sought energy security
- Promoted a cluster approach
- Developed a Green Economy Strategy Framework
- Wanted to take advantage of electricity crisis

Clusters address the limitations of, and tensions between, the public and private sectors

Environmental focus

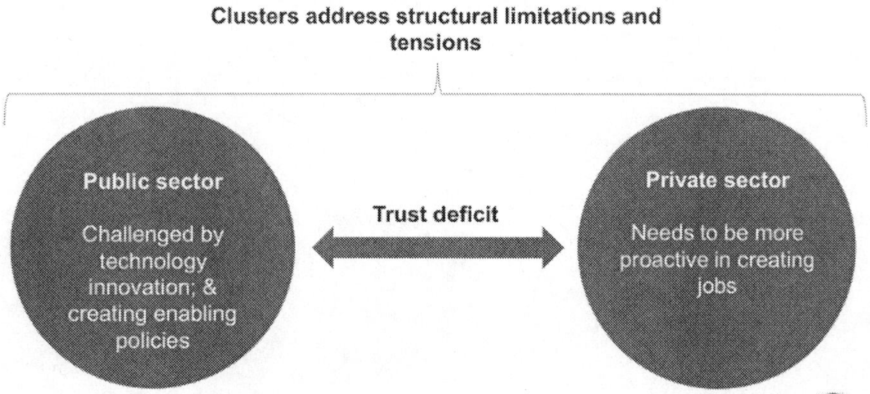

Green Cape is an economic development agency……

with a strong sustainability focus

KPIs: investment & job creation

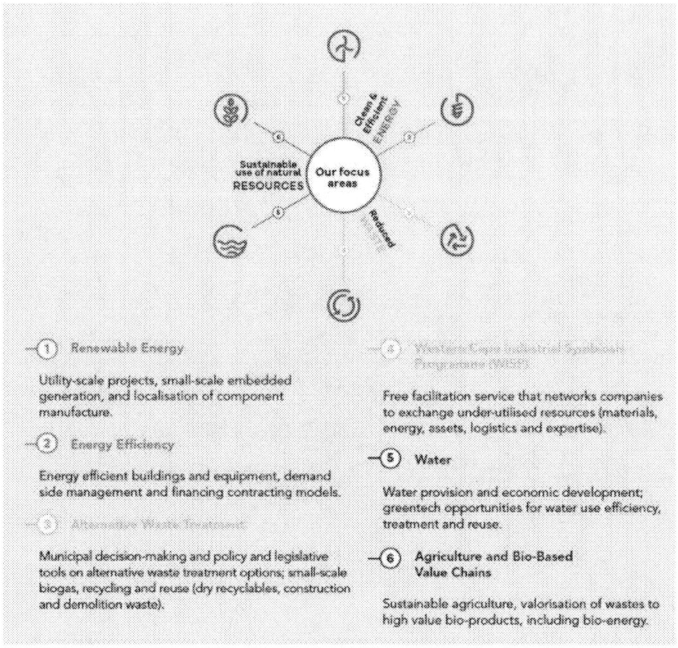

Green Cape activity

PASSIVE / INTERNAL ⟷ ACTIVE / CLIENT-FACING

Research:

- Field data through key relationships, site visits and networking
- Market Intelligence Reports
- Knowledge pieces
- Industry updates

Tools:

- Green-Agri Portal
- Decision-making tools e.g. waste, non-revenue water
- Regulatory and legislative advice e.g. tariff work
- Green Finance Database

Stakeholder engagement:

- Networking events
- Thought leadership presentations
- Setting up cross-sectoral and triple helix meetings
- Support for industry gatherings

Front-line activity:

- Proposed Special Economic Zone for Greentech manufacturing
- Western Cape Industrial Symbiosis Programme (WISP)
- Market Connect and Finance Lab

GreenCape | 8

GreenCape has over 1100 members, 60% of which are businesses and independent consultants

Total GreenCape members

1178

GreenCape members sign up to receive newsletters and stay informed on recent developments and networking events

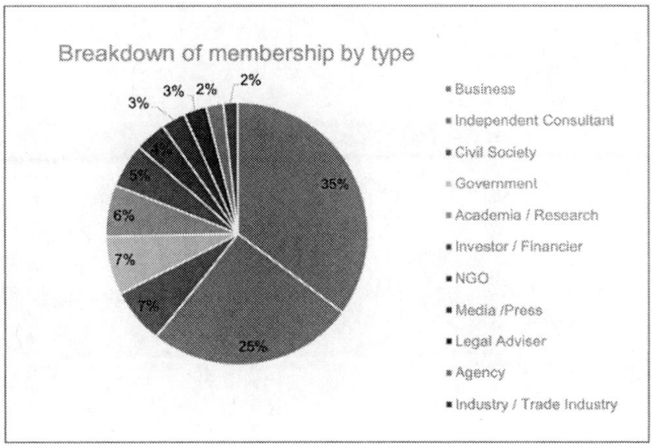

Source: GreenCape

GreenCape funding from all tiers of government + grants

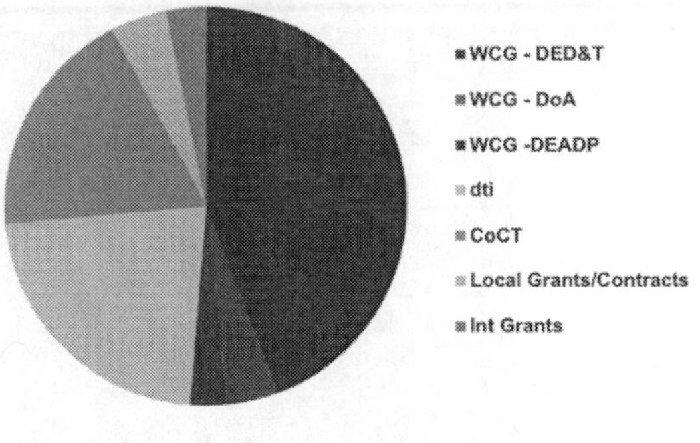

Status quo

- Western Cape Government Department of Economic Development and Tourism provides a core base of funding
- Funding planned over 3 year periods, but allocted annually
- Key metrics for funders are impact in terms of investment and job creation

Going forward

- Large multi-year funded projects such as *World Bank Market Connect and Finance Lab* form part of a movement towards sustainable funding

Green Cape achievements

- **ZAR 17bn investment** in the Western Cape since 2011
- **> 10,000 local jobs** created
- **> 1,000** members
- **1st African Cleantech** cluster invited to join the world leading regions – International Cleantech Network (ICN)

Example: Renewable Energy Independent Power Producers Purchasing Programme (REIPPPP)

- SA's most successful PPP since achieving democracy
- GreenCape's role was to act as a free and effective turnkey solution
  - Worked with Industry Associations
  - Attracted developers
  - Pushed through necessary troubleshooting

Example: Leading achievements at a national and international level

SKILLS DEVELOPMENT

INVESTMENT PROMOTION

Example: Green Fund Partners:

The World Bank Group

The objective of The World Bank Climate Technology Program (CTP) is to accelerate the growth of local green technology sectors that contribute to climate change mitigation and adaptation. **The CTP provides technical assistance and global insights to the Green Outcomes Fund.**

WWF South Africa

WWF-SA is part of the Worldwide Fund for Nature (WWF) global network, working to reduce humanity's footprint on the environment. **WWF-SA brings deep insights of local and global climate change mitigation as well as a green impact investing to the Green Outcomes Fund. WWF-SA will act as the GOF administrator during the initial implementation phase.**

UCT GSB Bertha Centre

The UCT GSB Bertha Centre is the first academic centre in Africa dedicated to advancing social innovation and entrepreneurship, especially through innovative finance. **The Bertha Centre has led the development of the Green Outcomes Fund and will continue to act as a local knowledge partner.**

GreenCape

GreenCape is a sector development agency supporting businesses operating within the green economy in South Africa. **GreenCape brings deep insights of the local green economy to the Green Outcomes Fund.**

Green Cape aims for a ruthless prioritization approach which is firmly adhered to

Illustration: Water Programme, 2017 - 2019

· · · · GreenCape focus threshold (note the slope)

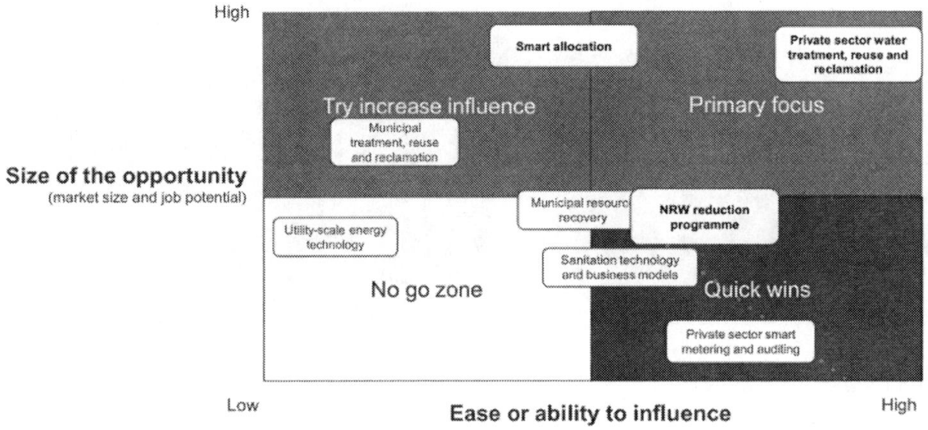

| 15

LEARNINGS

Generalising from Green Cape Experience

- **Successful Project Development requires**
  - **Clear articulation of project goals**
  - **Thorough understanding of related business models and practical business requirements**
- **Continued financial and institutional support from all tiers of government.**
- **Build trust and credibility by:**
  - **Alignment with national / provincial / local policies and strategies**
  - **Political neutrality**
  - **Legal independence**
  - **Technical competence**
- **Smart growth**
  - **Build on successes to drive expansion**
  - **Robust problem-solving approach**
  - **Impact orientation (ruthless prioritisation)**

GreenCape | 16

Positioning, People and Prioritisation drive Performance

Thank You

Lauren Basson and Mike Mulcahy

lauren@green-cape.co.za; mike@green-cape.co.za

www.green-cape.co.za

CHINA DOES NOT LACK WATER
SUSTAINABLE WATER USES AND MANAGEMENT MEETING PRIORITY GOALS OF THE UN 2030 PARIS AGENDA

PROFESSOR WALTER W. LOO
Ambassador of Peace (IAEWP UN NGO)

PROFESSOR MICHAEL Y. LUK PhD
CHAIRMAN ENVIRONMENT & TECHNOLOGY SERVICES (HONG KONG)

PROFESSOR I-SEN WANG
GROUNDWATER HYDROLOGIST GEOPHYSICIST

13126689981@163.com

INTRODUCTION

China is the largest developing nation on earth and is taking the lead in the compliance of United Nations 2030 Paris Agenda Sustainable Development 17 Priority Goals.

This paper analyses on the available water resources on the development of China. In general, people believes China lacks water due to explosive population growth.

Four thousand years ago, China's famous Emperor Yu was a hydrologist but his main duty was to conquer flood. Since Emperor Yu time, China has many good geologist but few are good in hydrology. The few hydrologists that are qualified Fellows of Chinese Academy of Sciences passed away in the last few years. These hydrologists are good hydrologist in defining the water resources but are not known to management massive water pollution caused by industrialization. The poor water quality management caused mostly by ignorant chemical and industrial engineers without cross disciplinary training for the last 30+ years.

Our author is trained in the western education system in the late 1960s and early 1970s under Dr. Madhi Hantush and the reputable New Mexico Tech at Socorro, New Mexico-----the 4th generation of groundwater hydrologist. We gain experience as consultants in the environmental fields with diverse multi-disciplinary practices and not through research.
China as a major developing nation is committed to the new United Nations 2030 Paris Agenda 17 priorities development goals established below [1].

International Association of
Educators for World Peace
和平大使奖
AMBASSADOR OF PEACE AWARD
This is to certify that
Walter Loo
罗怀涛
Has been confirmed by the
Secretariat's Executive Council
Ambassador of Peace
联合国环保和平大使
In recognition of his outstanding dedicated
service to the cause of social justice and world peace.

UNITED NATIONS 2030 PARIS AGENDA
17 SUSTAINABLE DEVELOPMENT GOALS
联合国2030巴黎议程
17可持续发展目标
GOAL 目标 1: NO POVERTY 不贫穷
GOAL 目标 2: ZERO HUNGER 零饥饿
GOAL 目标 3: GOOD HEALTH AND WELL-BEING
健康和幸福
GOAL 目标 4: QUALITY EDUCATION 素质教育
GOAL 目标 5: GENDER EQUALITY 两性平等
GOAL 目标 6: CLEAN WATER AND SANITATION
清洁用水和卫生
GOAL 目标 7: AFFORDABLE AND CLEAN ENERGY
负担得起的清洁能源
GOAL 目标 8: DECENT WORK AND ECONOMIC GROWTH
体面工作和经济增长
GOAL 目标 9: INDUSTRY, INNOVATION AND INFRASTRUCTURE
工业、创新和基础设施
GOAL 目标 10: REDUCED INEQUALITIES 减少不平等
GOAL 目标 11: SUSTAINABLE CITIES AND COMMUNITIES
可持续的城市和社区
GOAL 目标 12: RESPONSIBLE CONSUMPTION AND PRODUCTION
负责任的消费和生产
GOAL 目标 13: CLIMATE ACTION 气候行动
GOAL 目标 14: LIFE BELOW WATER 水下生活
GOAL 目标 15: LIFE ON LAND 陆地上的生活
GOAL 目标 16: PEACE, JUSTICE AND STRONG INSTITUTIONS
和平、正义和强有力的机构
GOAL 目标 17: PARTNERSHIPS FOR THE GOALS
实现目标的伙伴关系
SUSTAINABLE
DEVELOPMENT
GOALS

CRITICAL REVIEW OF WATER MANAGEMENT AND WATER AVAILABILITY IN CHINA

Due to the management structure of the government management, the water resources decision makers often are without technical background and therefore usually follow other western countries water resource decisions and without innovations.

Due to the severe poor management of water resources for the last 30+ years, severe water pollutions in surface and groundwater occurrences were reported above.

We have observed the following phenomenon that can provide improvements by newer technologies on the following projects:

Water conservation in wastewater treatment

Advanced seawater desalination technologies

Advanced water desalination from seawater, salt-alkali land and wastewater

Available water moisture in soil (vadose zone) as a water supply

Evaluation of the large fresh groundwater resources underlying the northwest lowlands, highlands and mountains of China

NEW WATER SOURCES

HEIBEI PLAIN AVAILABLE WATER RESOURCES

The Hebei Plain area including Beijing, Tianjin and Hebei Province is known for lack of traditional freshwater resources.

However, our understanding of our hydrosphere has broadened. Many new fresh water sources are readily available with simple cost effective treatment and recovery systems. These new water sources are listed as follows:

SOLAR DISTILLATION OF SEAWATER

Solar distillation of Bohai seawater by solar heating. With solar electricity, the tail turbid yellow seawater can be treated and discharged as warm clear seawater economically (less than RMB$5 per ton). As a result, the 300 KM coastline will become highly valuable real estate properties have warm clear blue water beaches (Figure 1).

SALT ALKALI FLAT LAND

Shallow saline water underlying the salt-alkali flats along the coast can also be converted into distilled water by solar heating which can be used as soil cleaning and nano bubble irrigation water under enclosed greenhouses with extreme water and energy conservation. The nano bubble water (RMB$10 per ton) will eliminate the use of conventional fertilizers and pesticides while double or triple the GDP per acre of greenhouse (Figure 2).

ZERO WASTEWATER DISCHARGE

All the wastewater treatment plants discharge in the area should be banned. This should promote the principles of Chairman's proclamation as presented in his book published in October 2014—The Principles of Ruling China. Specifically, he claimed to fulfill the goals of passing on to many future generations Blue Sky, Green Earth, and Clear Water on July 18, 2013 (page 211-212). This is the basically the goals of "Zero Discharge and Emission into the Atmosphere, Soil and Water". The wastewater discharge today is not complying with Chairman Xi's goals. Yes, it can be done and also create high economic return. The wastewater treatment plant tail water can be converted back into distilled water by solar distillation within an enclosed environment at less than RMB$5 per ton. The cost of new water from the South to North Water Reclamation Project to Beijing cost RMB$12 per ton. There is profit to be made and it will comply with Chairman Xi's Zero Discharge Principles. (Figure 3)

RECOVERY OF WATER FROM SOIL MOISTURE IN VADOSE ZONE

The ground underlying the Hebei Plain or Basin including about 400 meters of Quaternary fanglomerate sediments. On the average, there are about 50 meters of unsaturated soil (vadose zone) which contain 20 to 30% moisture content. Of which about 10% of the soil moisture can be recover easily by solar hot air injection and recovery of humid warm air and then condensed as distilled water. The cost of recovery varies from RMB$5 to $10 per ton depending on the air permeability of the unsaturated soil. It is estimated that with 40 meters of moist soil and 10% moisture content that can be recovered as distilled water, there is enough water to supply a population of 100 million that can use 25 gallons (100 liters) per day per person for 400 years (Figure 4 & 16) .

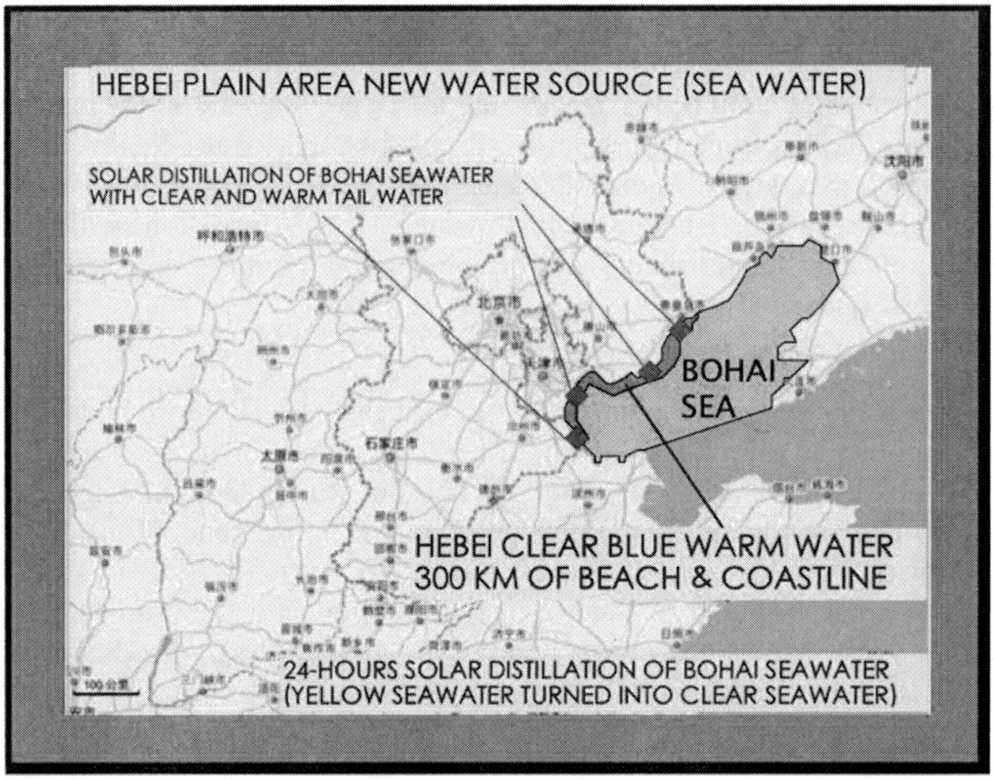

FIGURE 1 ENCLOSED SOLAR SEAWATER DESALINATION

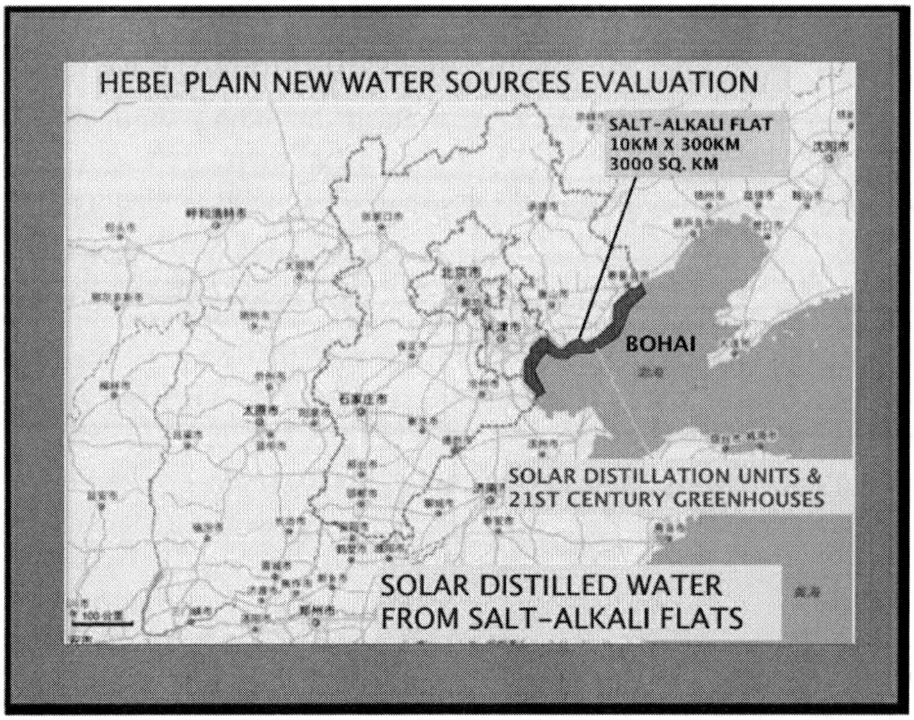

FIGURE 2 ENCLOSED SOLAR SEAWATER DESALINATION OF SALT ALKALI LAND

FIGURE 3 ZERO DISCHARGE WASTEWATER MANAGEMENT

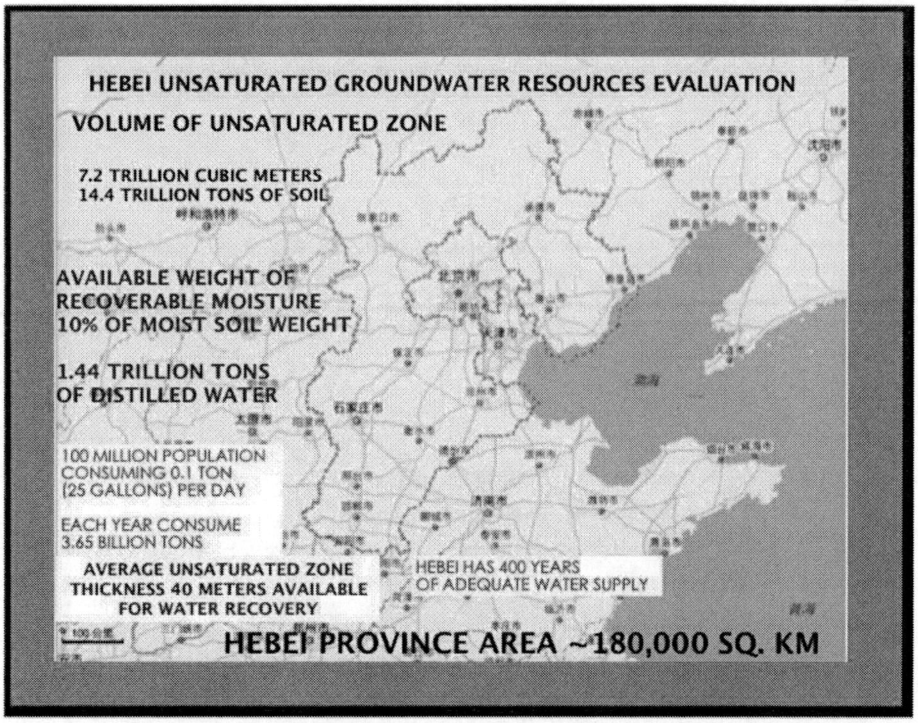

FIGURE 4 WATER FROM SOIL MOISTURE

HYDROLOGICAL ANALYSIS OF THE HEBEI PLAIN WATER RESOURCES THAT WILL MEET UN 2030 PARIS AGENDA PRIORITY GOALS :

There are 2 main water resources features that supply water to the Hebei Plain area; groundwater and the South-to-North Water Diversion Project.

The groundwater withdrawal basically has caused aquifer depletion for the last 40 years of pumping. The shallow groundwater is badly contaminated.

The North-to-South Water Diversion is delivering about 2.2 billion tons of Class II water quality per year, Figure 5.

Worst of all, numerous wastewater treatments plants of Beijing, Tianjin and Zhijiazhong have been discharging Class IV and Class V black stinking water into the existing wastewater waterways polluting natural waterways and polluting the shallow ground water.

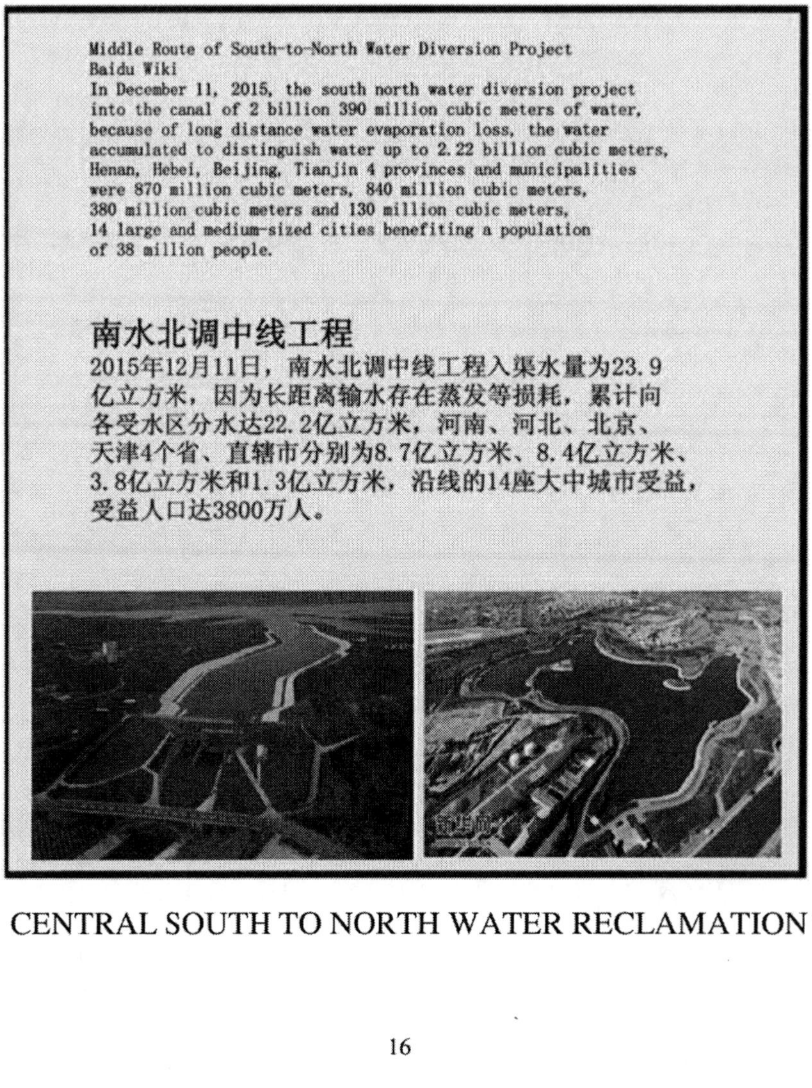

FIGURE 5 CENTRAL SOUTH TO NORTH WATER RECLAMATION PROJECT

Recommendations to meet UN 2030 Paris Agenda Priority Goals:

Immediate treatment of poor water quality discharged wastewater by solar desalination. The total discharge is estimated at 6 million tons per day or 2.2 billion tons per year.

Enclosed solar desalination plants can be set up along the northern banks of existing waterways (facing south) Figure 6 and 7.

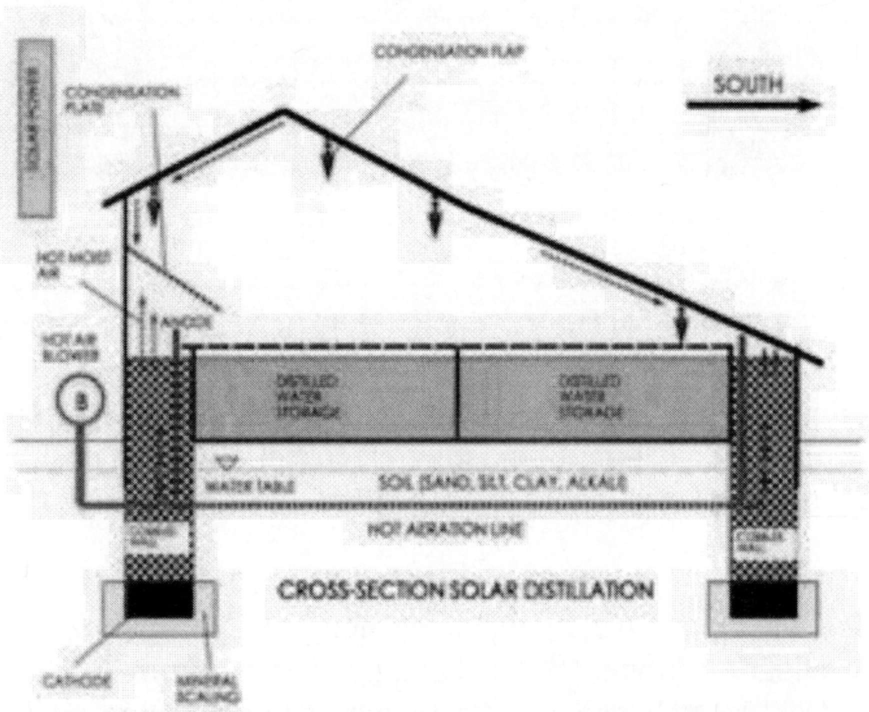

FIGURE 6 SOLAR SEAWATER DESALINATOR

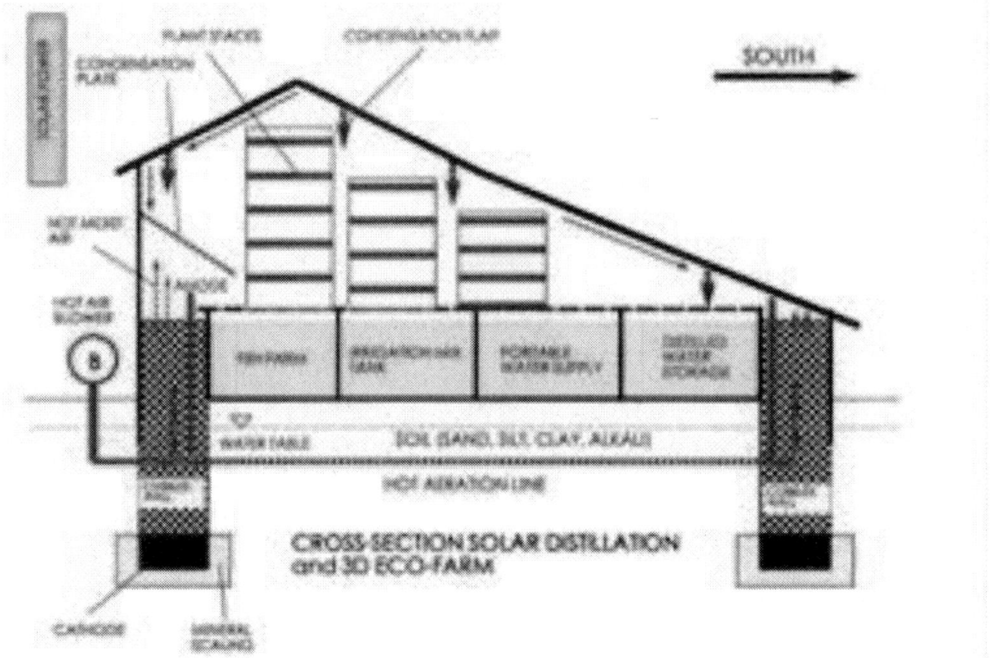

FIGURE 7 21st CENTURY AGRICULTURAL GREENHOUSE

This will replace the water delivery from the South-to-North Water Diversion Project. The saving of existing electrical pumping cost (fossil fuel power plant emission) will meet UN 2030 Paris Agenda air emission priority goals.

The increase in future water requirements will be met with enclosed seawater desalination along the salt-alkali land and from Bohai seawater. There should be enough capacity to recharge the desalinate Class I type water into the shallow aquifer.

The increase water demand from agricultural irrigation can be recovered from soil moisture in the existing vadose zone soil Figure 8, 9 and 10. Or, irrigation water can be supplemented by enclosed salt water desalination.

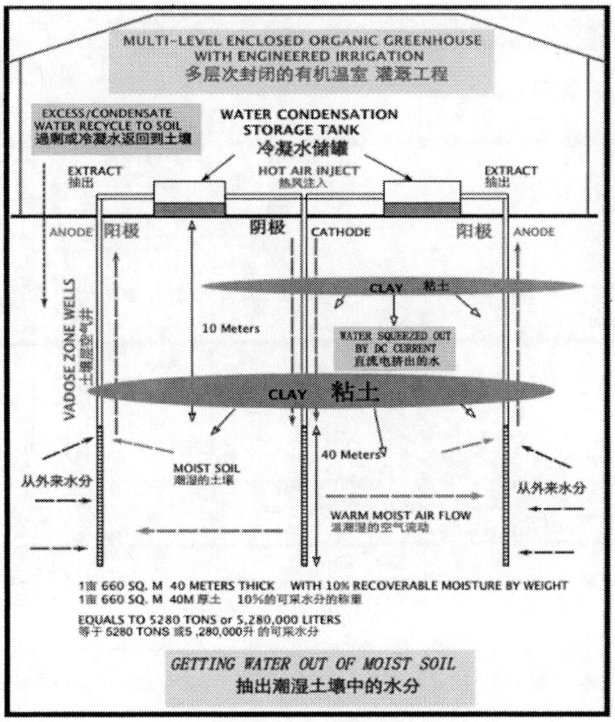

FIGURE 8 GETTING WATER OUT OF MOIST SOIL

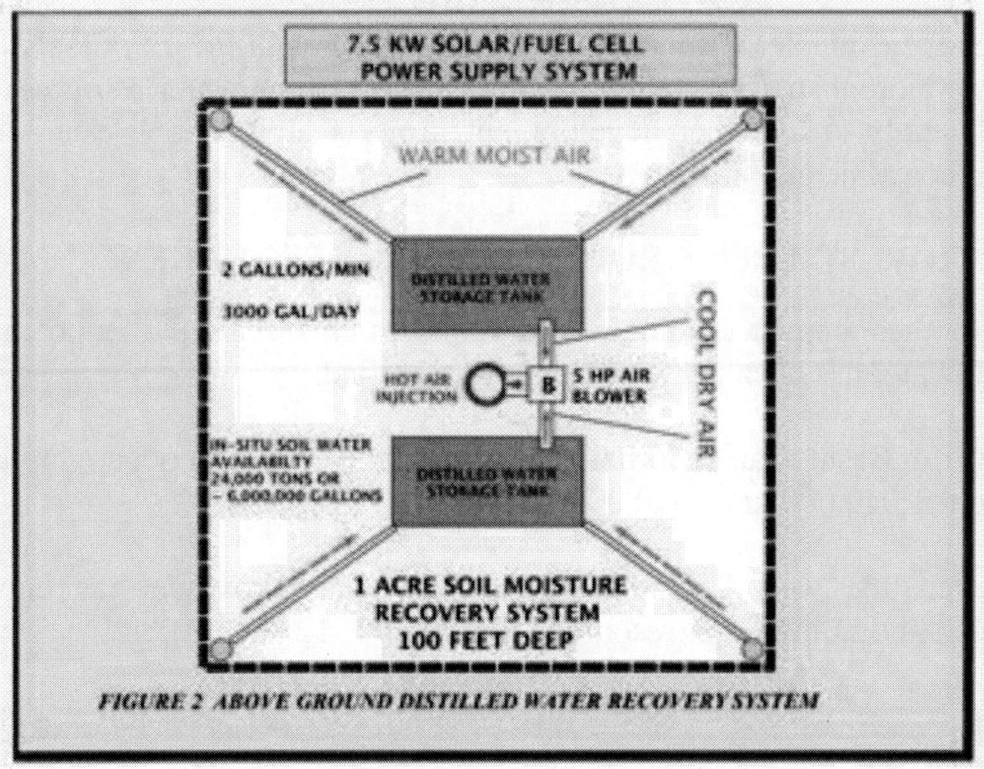

FIGURE 9 LAY OUT OF CONDENSATE WATER COLLECTOR

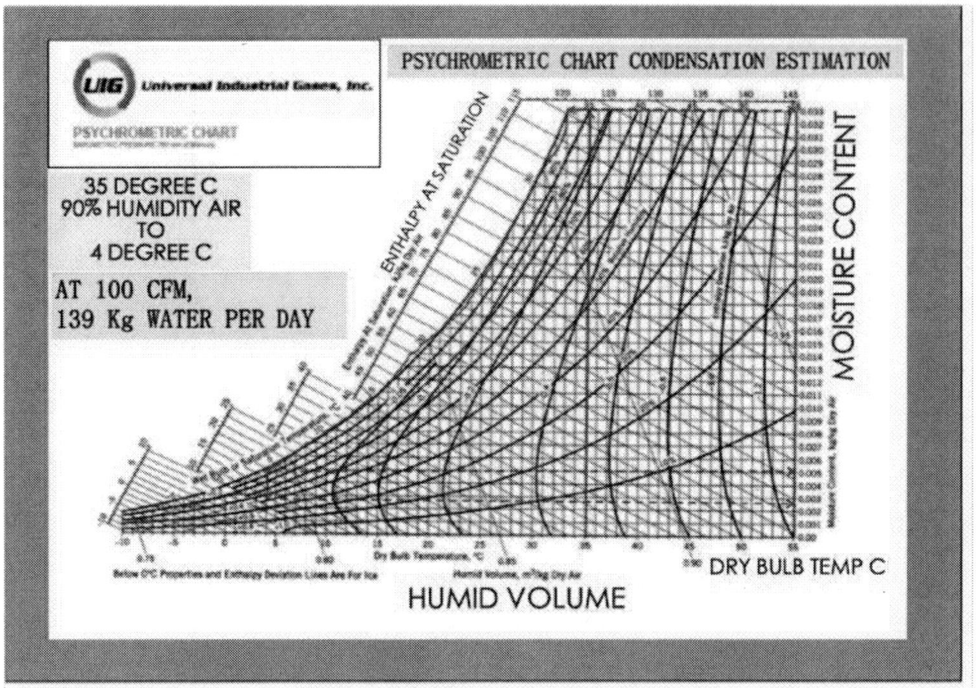

FIGURE 10 PSYCHROMETRIC WATER CONDENSATION CHART

The above example will meet UN 2030 Paris Agenda Priority Goals [1] and provide more than sustainable development for China water resources. Also, it will apply to all other development projects along the Western Pacific Ocean coastline of China.

NORTHWEST AREA GROUND WATER RESOURCES

Due to the complex geologic conditions in this region, the potential of groundwater has been neglected or not quantified.

The Tarim Basin is an endorheic basin in northwest China occupying an area of about 1,020,000 km^2 (390,000 sq mi).

However, the vast Tarim Basin of Xingjiang Province has enormous brackish and saline water resources (Figures 11, 12 and 13).

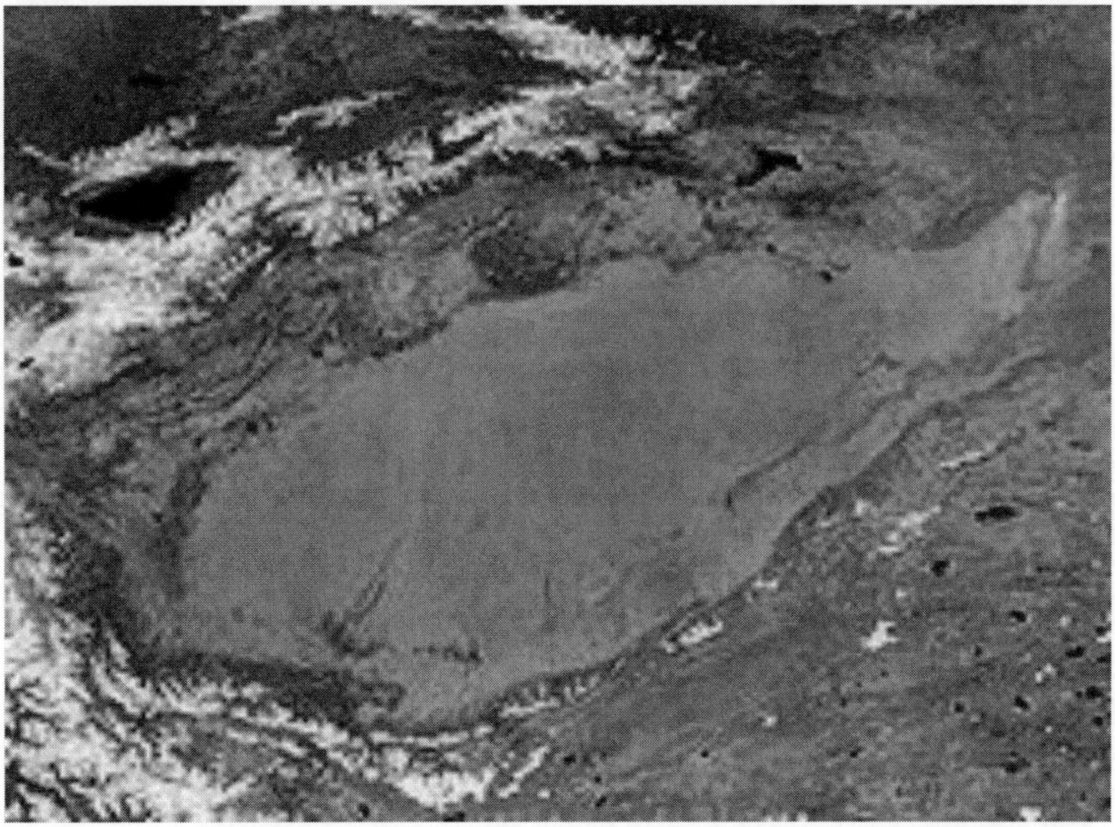

FIGURE 11 TARIM BASIN XINGJIANG PROVINCE CHINA

FIGURE 12 BRACKISH AND SALINE WATER OF TARIM BASIN

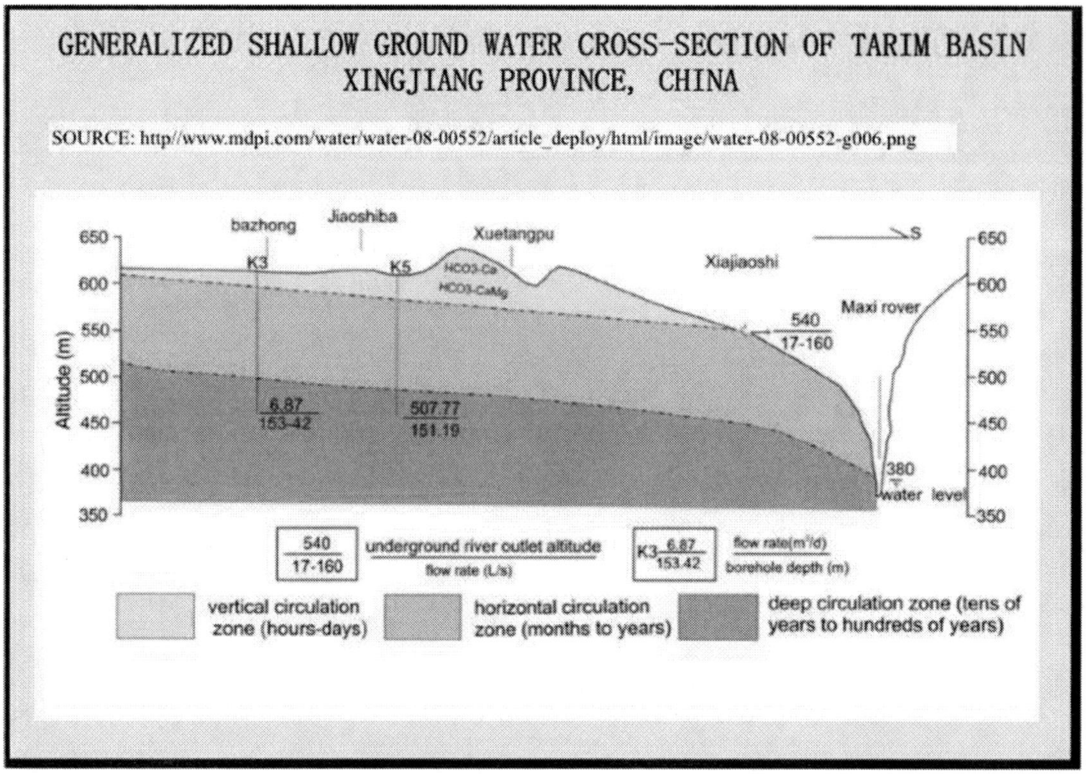

FIGURE 13 GENERALIZED SHALLOW GROUND WATER CROSS-SECTION OF TARIM BASIN XINGJIANG PROVINCE, CHINA

The saturated thickness of the sedimentary rocks is at least 10,000 meters and the depth to ground water is from surface to about 100 meters.

Assuming that the developable thickness of ground water is 2,000 meters and with an effective porosity of 5%, the volume of developable groundwater for saline water desalination is estimated at 100,000 billion cubic meters or tons.

The average annual water consumption of China averages about 600 billion tons.

The Tarim Basin available groundwater for development is estimated at least 160 years without conservation and without recharge from the surrounding Tibetan Plateau and Tianshan Mountains.

The groundwater pumping can be deployed by periodic solar power assist groundwater pumping as shown in Figure 14.

FIGURE 14 CYCLIC SOLAR GROUNDWATER PUMPING OF TARIM BASIN

Proposed More Than Sustainable Development Technologies Goals (UN 2030 Paris Agenda) Economical Feasibility Evaluation For Tarim Basin

[2] Loo et al, 2016, Key Note Speech IRSEC-2016, 4 Dimensional 24 Hours Thermal Solar Energy And 21st Century Agricultural Development, Morocco

The proposed 4 Dimensional 24 Hours Thermal Solar Energy [2] is a must to alleviate the poverty in China. Conventional electric power plants will take at least 2 years to setup which will not be able to meet the SDGs of the UN 2030 Paris Agenda. The proposed solar energy will be demonstrated in China within 1 year (to be financed by OBOR AIIB) and the technology will be transferred to the needy part of China and start manufacturing along the coastal areas and the Northwestern China. Cost of electricity to these areas will be less than US$0.015 per KWH. This solar system can produce both thermal energy and electricity for use by the community and create employment opportunities (Figures 11 to 18).

The solar energy is a must to desalinate seawater [3} along the coastal regions and the dry region of China. This is both water supply to the local population and also vital irrigation water supply (Figures 11 to 18).

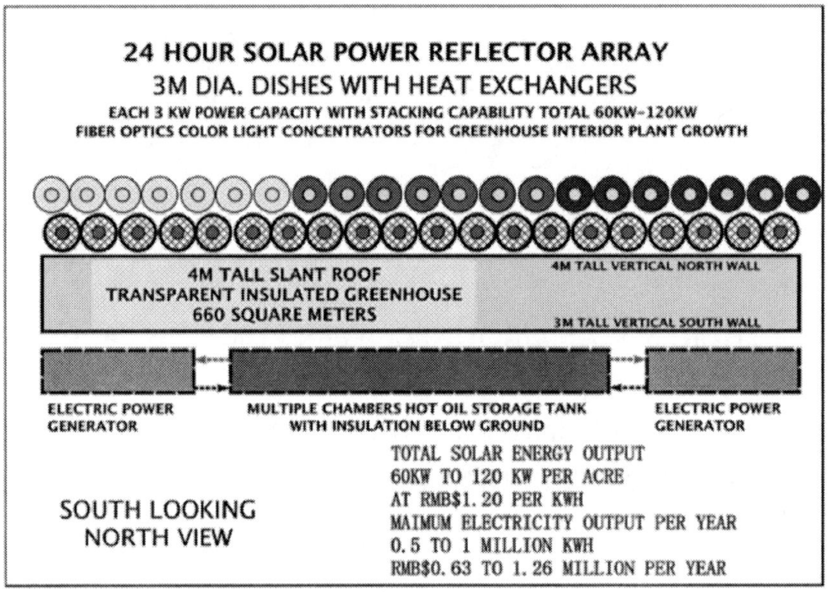

120+KW 24 HOURS SOLAR POWER ON 660 SQ. METERS
RMB$800,000+ SOLAR ELECTRICITY PER YEAR

FIGURE 15 4 DIMENSIONAL 24 HOURS SOLAR SYSTEM

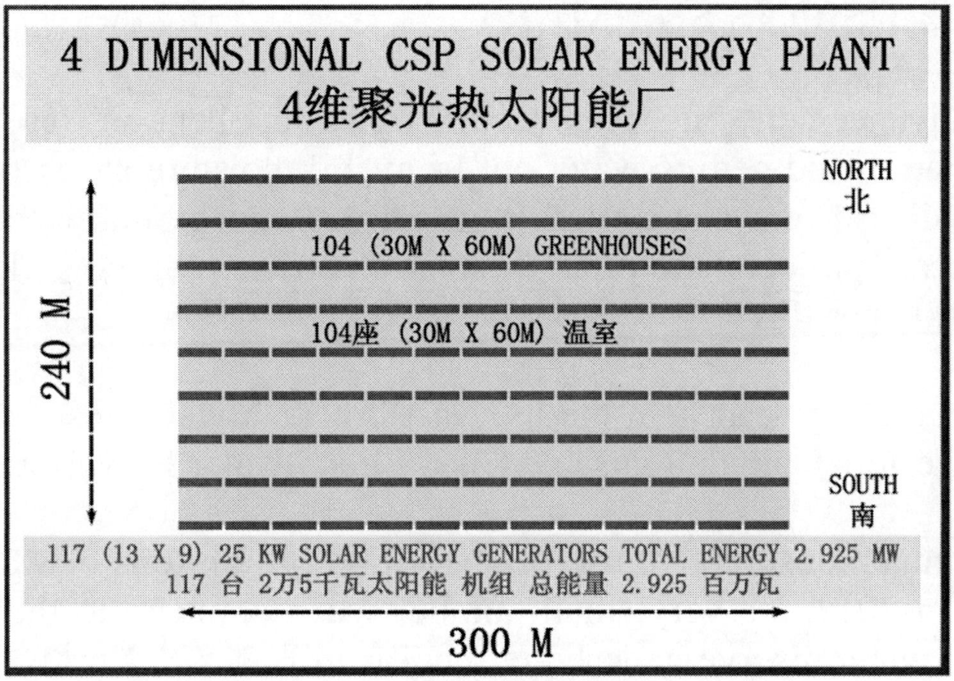

FIGURE 16 18 ACRES 4 DIMENSIONAL CSP POWER PLANT & 21ST CENTURY ORGANIC GREENHOUSE SYSTEM 2017

The 18 acres 4 dimensional CSP Power Plant can generate about 3 MWs or RMB 30 million per year (if the price is guaranteed to be RMB 1.20 per KWH) or RMB$13 million per year (if the electricity price is RMB$0.50 per KWH).

The investment per watt is estimated at RMB$ 7.8.

The total investment is estimated at RMB$ 23,400,000.

The return on investment is estimated to be less than 2 years.

The life expectancy of the power plant is 15 years, and the electricity price is estimated at RMB$ 0.06 per KWH.
It's just a 100 acre power plant.

If the power plant is greater than 10000 acres, then the cost of investment per watt will be estimated to be less than RMB$ 4.

Electricity price is estimated to be less than RMB$ 0.03 per KWH.

CONCLUSIONS SUMMARY

The advancement in solar technologies application on water desalination and getting water out from soil moisture at reasonable price will rid of water pollution in surface and groundwater. For large scale projects, the price of solar electricity can be as low as US$0.004 per KWH.

Also, the above can help the use vast brackish water resources in Northwestern China into usable Class I good water for various uses.

The annual consumptive use of about 600 billion tons of water in China in 2012 is a very small number when compare to the fresh ground water reserve underlying the Tarim Basin of Northwestern China.

Also, the evaluation on sustainable water resources is economically feasible and will meet the UN 2030 Paris Agenda [1] Priority Goals and meeting the zero emission and zero discharge goals [3].

Based on the above analysis and estimations, china does not lack water.

Also, a separate evaluation paper of the available groundwater resources (Third Pole Aquifer) underlying the Tibetan Plateau of China which will be published in 2018.

In conclusion, China does not need the Surface Water Diversion Project from Tibet Plateau to Xingjiang which was proposed in early 2017. It is a waste of monetary resource with the proposed project value posted on Baidu internet exceeding US$200 billion or about RMB 1500 billion Yuan.

REFERENCES CITED

[1] UN 2030 Paris Agenda, 2015 & SDGs Adoption At G20 September 2016 Summit Conference, Hangzhou, China

[2] Loo et al, 2016, Key Note Speech IRSEC-2016, 4 Dimensional 24 Hours Thermal Solar Energy And 21st Century Agricultural Development, Morocco

[3] Loo et al, 2015, Zero Water Discharge Management, World Academy Of Sciences, Melbourne, Australia

Sustainability education approaches for life-long learning

Are we there yet?

Prof Jarka Glassey and Dr Sue Haile
School of Engineering, Newcastle University
Jarka.glassey@ncl.ac.uk

Bac

- Ing STU Bratislava (MEng b
- PhD Newcastle University (b
- Professor of Chemical engin

Engineers

Forum (pedagogical

Context

- Importance of sustainability teaching within chemical engineering degree courses widely recognised
- Contained in various mission statements both for education and business/company operation
- Requirement of various accreditation processes

Education challenges

- How do we prepare the students for the future challenges?
- How do we ensure 'quality experience' for the increasing student numbers?
- Not just doing the 'same old' numerou

arter' things

xpectations,

most importantly....

- How do we ensure our students gain the right attitudes and skills to carry on learning throughout their professional careers?

Some examples

- Some IChemE initiatives in supporting sustainability teaching
  - Sustainability Teaching award
  - Education SIG workshops

- Examples of sustainability teaching at UG level
- Examples of sustainability teaching at PG level

Awards and prizes

- **Macnab-Lacey Student Design Prize** –awarded to the design project that best shows how chemical engineering practice can contribute to a more sustainable world.
- In order to increase impact, a joint **Sustainability and Education SIG Sustainability Teaching Award** for individuals or teaching teams
- encourage teachers to think of sustainability as a key element of lecture courses
- influence CE Departments to position sustainability at the heart of the curriculum
- provide a showcase for teaching talent and reward achievement in the field of sustainability education

Sustainability Teaching Award

- **2015 winner UCL** commended for utilising scenario based learning in the first year to introduce sustainability principles in the absence of a significant body of chemical engineering knowledge and building and strengthening these concepts through the course leading up to the design project

UCL

- **2015 Highly commended Newcastle University** for utilising case studies and workshops to teach the concepts, methods and principles of sustainability

Newcastle University

Sustainability Teaching Award

- **2016 winner UC Cork** commended integration of sustainability teaching across the curriculum, with good examples of interdisciplinary projects and varied assessment with student comments indicating their appreciation of the approach
- **2016 Highly commended University Manchester** for good integration of sustainability teaching throughout the curriculum with good design project case studies and the involvement of industry

EdSIG workshops

- Bring together academics, industrialists and students to discuss good practice, drivers and challenges in the embedding sustainability into the curriculum
- Industrial message at the last workshop
  - ensure strong technical competence and a broad sustainability overview, we will do the detailed training whilst in post
  - interdiscipinarity is key
  - communications
- Examples of academic practice embedding sustainability into UG curriculum
- Students' perspective on the effectiveness

Examples UG curriculum

- Integrated within the curriculum, progressively more complex topics covered, relying on more technical knowledge
- Year 1 – awareness raising, speakers from industry highlighting sustainability agenda, initial sustainability assessment of various known processes (e.g. citric acid, sugar production, etc)
- Year 2 – more technical topics covered in workshops (e.g. fracking, etc), role plays, stakeholder discussions, technical underpinning of sustainability decisions
- Year 3 – full sustainability assessment, LCA, environmental, safety, etc. design project

Resources

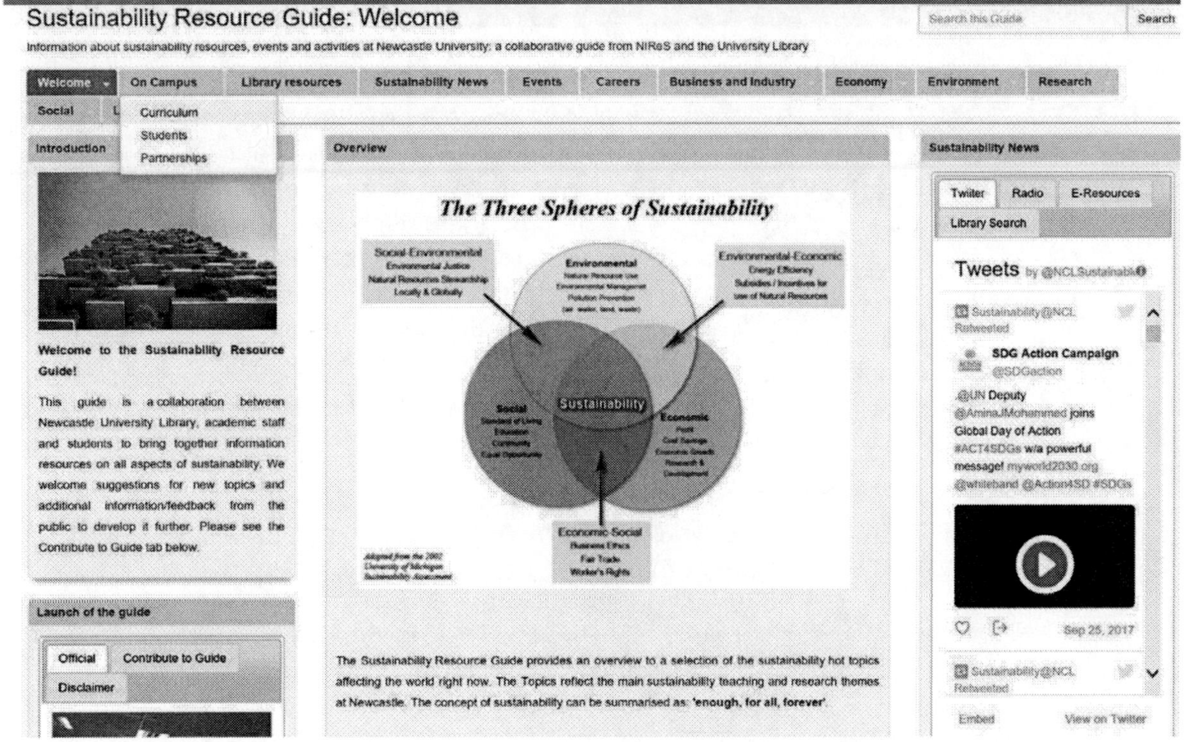

Examples PG curriculum

- interdisciplinary – mostly science background MSc and MEng Chemical engineers in mixed groups
- each group is part of a consortium trying to develop and raise funding for a new project eg. bioethanol plant , autoclave or gasifier.
- students identify the potential financial benefits, environmental impacts and societal implications and make an oral presentation.
- other students role play stakeholders having been given suggestions for questions as if at a public meeting

Student experience

"This exercise has increased my awareness of the lengths businesses and organisations must go to satisfy all stakeholder concerns when considering expanding or developin

"I have learnt that despite the best intentions and reasonable evidence, projects such as the one investigated can still be turned down due to the poor reputation stemming from … hensible activities … ss responsible … nies."

"I have learned to listen to everyone's ideas, as there are always things you would have never thought of and it gives a different perspective on how to go about a problem. "

Biomimicry

LimpiCook Intelligent cooking. Naturally.
Features
Design
Tech Specs
Buy Now
Say Hello to LimpiCook©
A revolutionary way to cook moist food in the microwave.
See how it works
CLICK TO PLAY
Intelligent Cooking, Naturally.
LimpiCook© is inspired by nature, to provide a clean, safe and efficient way to cook great food; even in the microwave.
See how we were inspired >

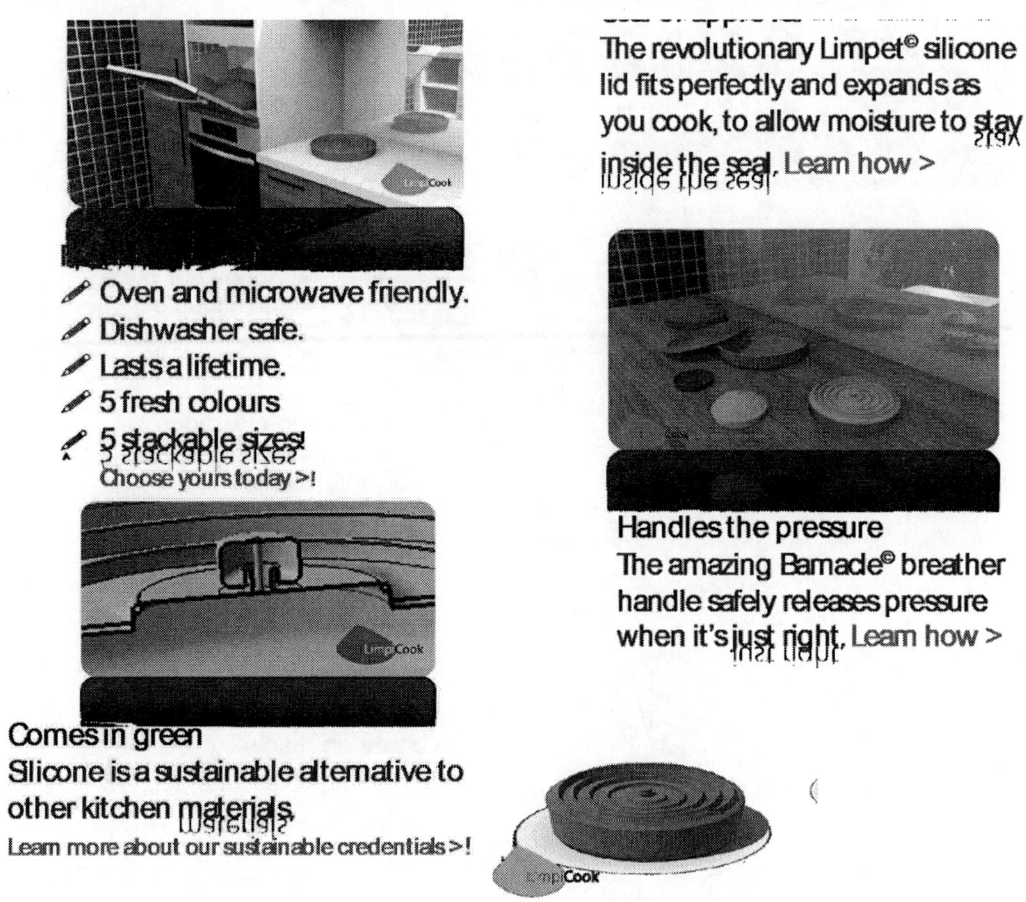

The revolutionary Limpet© silicone lid fits perfectly and expands as you cook, to allow moisture to stay inside the seal. Learn how >
Oven and microwave friendly.
Dishwasher safe.
Lasts a lifetime.
5 fresh colours
5 stackable sizes!
Choose yours today >!
Handles the pressure
The amazing Barnacle© breather handle safely releases pressure when it's just right. Learn how >
Comes in green
Silicone is a sustainable alternative to other kitchen materials.
Learn more about our sustainable credentials >!

Learning Outcomes

- Risk Assessment: the need to identify and manage the many types of risk which include societal, financial, economic as well as environmental and H&S risk. Supply chain auditing
- Effective stakeholder dialogue and communication strategies including which media to use

How is it assessed?

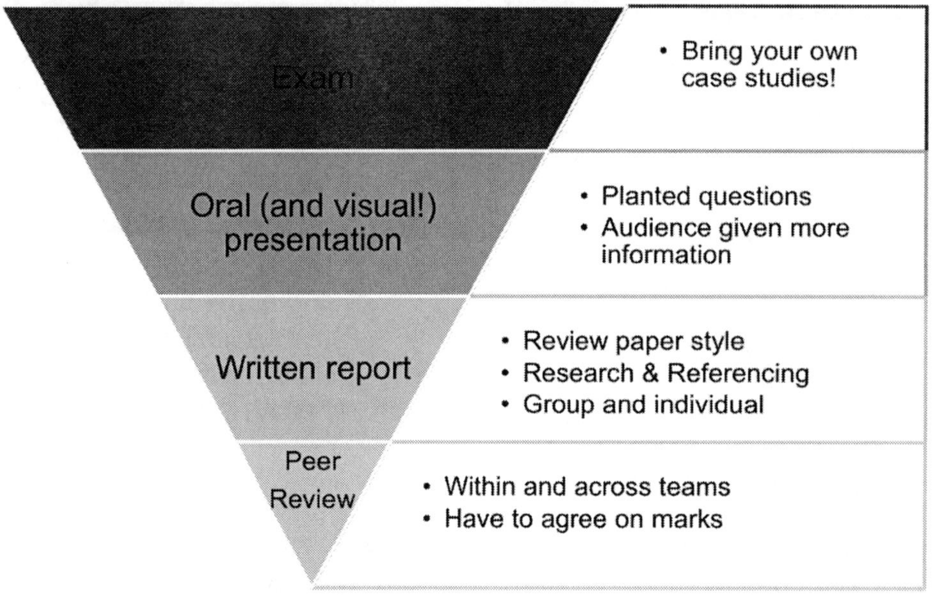

Learning by case study - Key Aspects

Role-play	Stakeholder POV	Regulatory	Research
Hard and soft skills	Multidisciplinary	Open Brief	'Real world' teamwork
Management	Field trip	Traceability	Peer review
	Self reflection	Media	

Concluding remarks

- Sustainability has to be embedded into the curriculum
- Working on real life issues, with industrial input important
- Strong discipline knowledge, but working in interdisciplinary teams
- Happy to share resources
- Share your good practice through ECE journal and apply for Sustainability Teaching award

Acknowledgements

- All colleagues involved in workshops and judging of the awards and prizes mentioned
- All my Newcastle colleagues involved in the delivery of the sustainability agenda

MANCHESTER
1824
The University of Manchester

Sustainable Industrial Systems

A decision-support framework for integrating circular economy principles into business practice

Joan Fernandez Mendoza, Alejandro Gallego Schmid and Adisa Azapagic

The University of Manchester

ICOSSE 2017, Barcelona

www.sustainable-systems.org.uk

Sustainable Industrial Systems

Circular economy concept

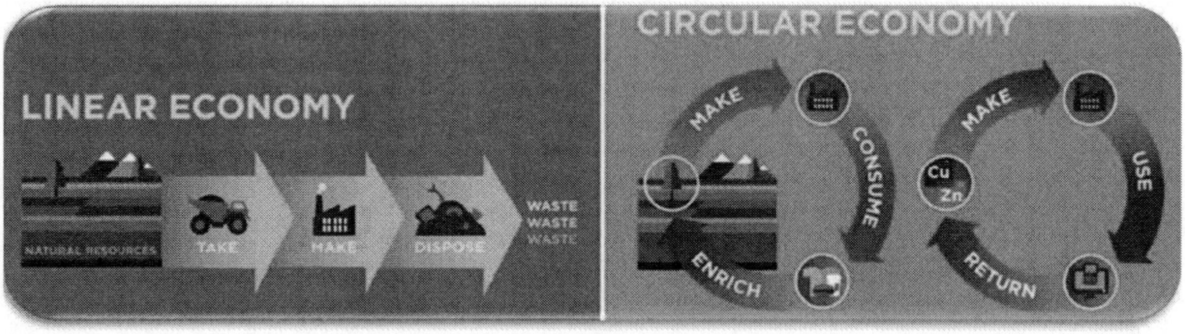

Image: EMF (2015)

www.sustainable-systems.org.uk

Sustainable Industrial Systems

CE approaches and tools

- Examples of approaches/frameworks
  - RESOLVE
  - Closed-loop supply chains
  - Product-service systems

- Examples of tools
  - Forecasting/backcasting
  - Business model canvas
  - Design for the environment
  - Life cycle sustainability assessment

www.sustainable-systems.org.uk

Sustainable Industrial Systems

RESOLVE

Regenerate	Restoring and regenerating natural capital
Share	Maximising asset utilisation
Optimise	Optimising system performance
Loop	Keeping products and materials in closed-loops
Virtualise	Dematerialising
Exchange	Selecting resources and technologies wisely

ELLEN MACARTHUR FOUNDATION
Rethink the future

www.sustainable-systems.org.uk

Sustainable Industrial Systems

Strategy for implementing CE in practice

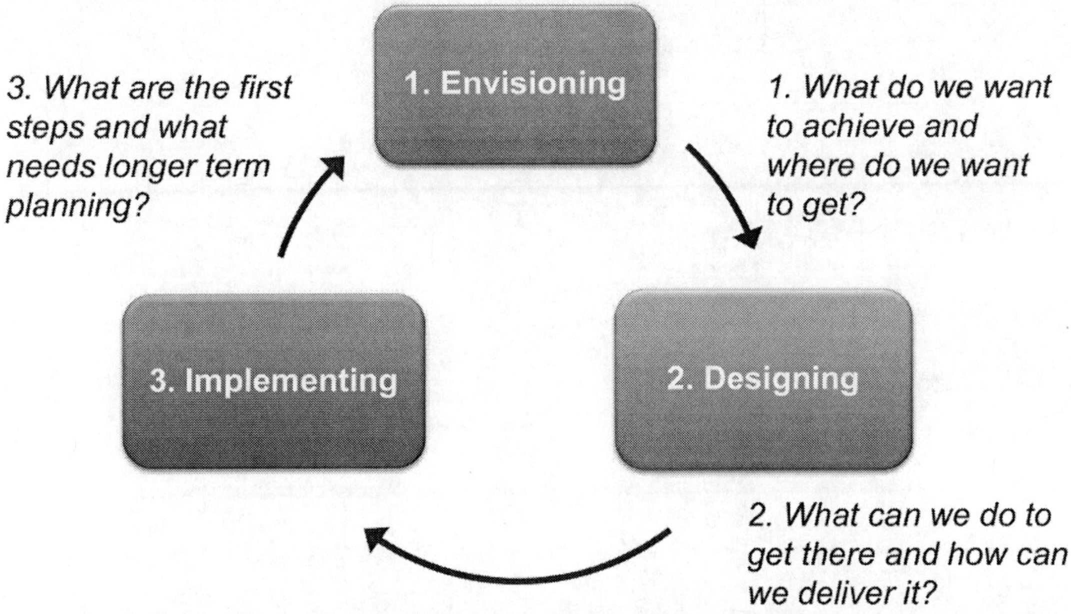

Mendoza et al. (2017) *J Ind Ecol 526-544.* www.sustainable-systems.org.uk

Sustainable Industrial Systems

Strategy for implementing CE in practice: BECE

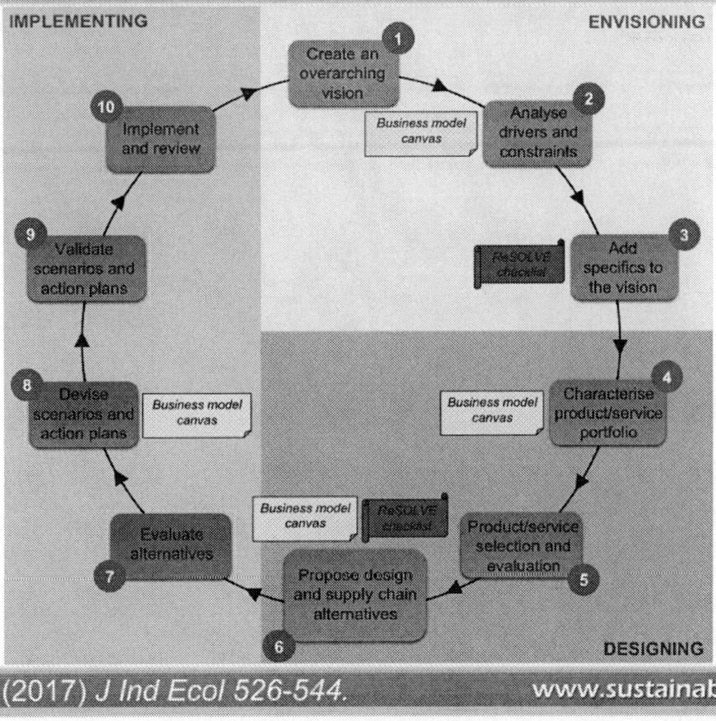

Mendoza et al. (2017) *J Ind Ecol 526-544.* www.sustainable-systems.org.uk

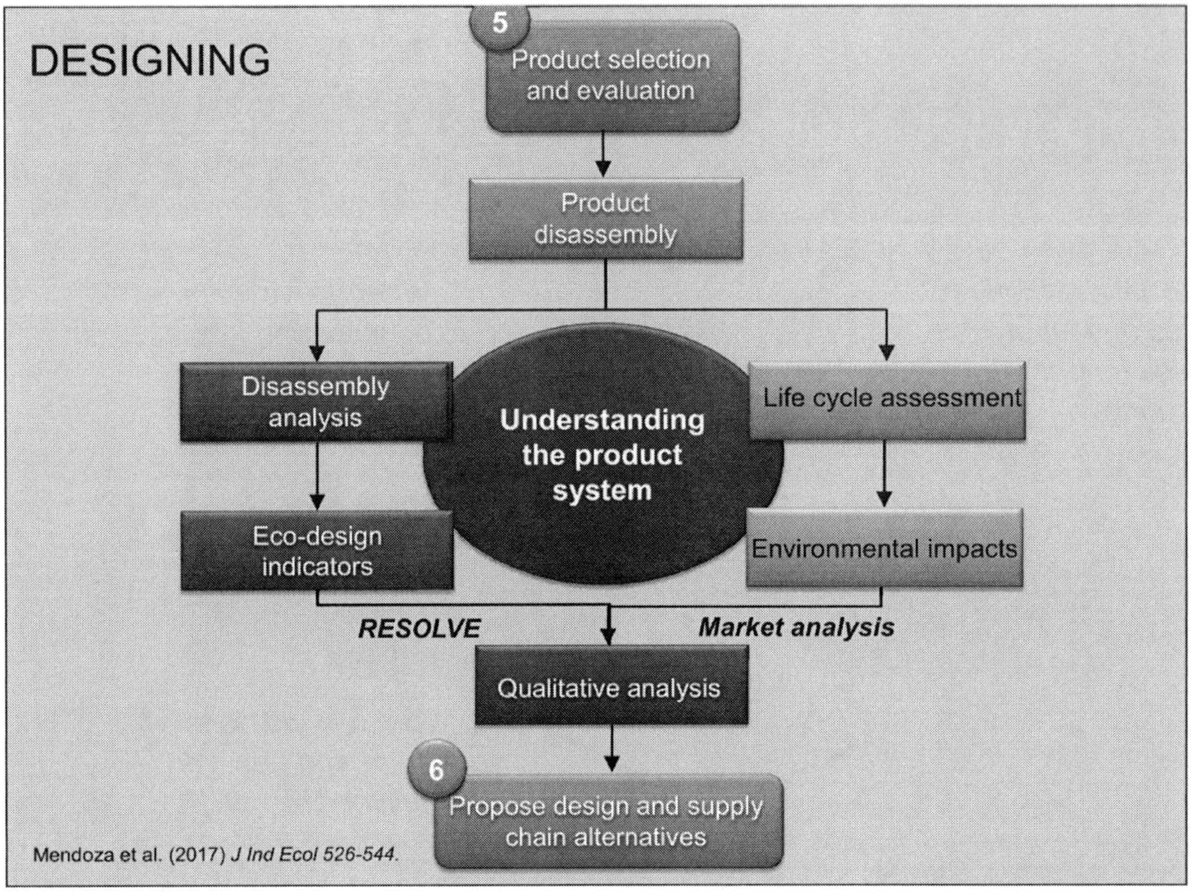

Sustainable Industrial Systems

An example product: Vacuum cleaner

www.sustainable-systems.org.uk

Sustainable Industrial Systems

Product disassembly

Gallego Schmid et al. (2016) *SciTot Env 559192-203.* www.sustainable-systems.org.uk

Sustainable Industrial Systems

Disassembly analysis

Product disassembly characterisation	Outcomes
Total number of pieces	150
Pieces theoretically not required	41
Number of task repetitions	313
Number of tool manipulations	116
Number of non-value-added tasks	104
Disassembly effectiveness	**21%**

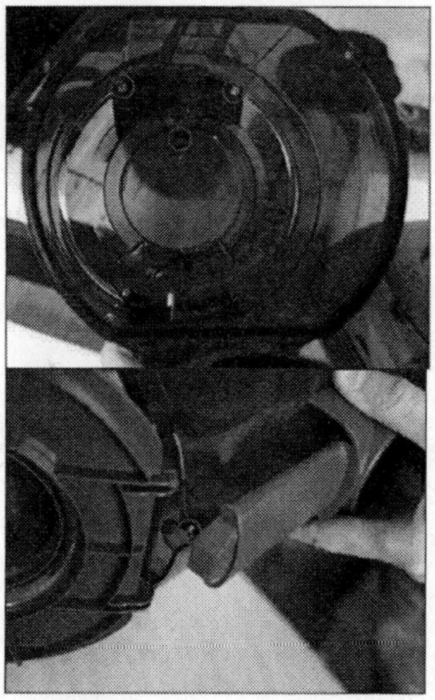

Mendoza et al. (2017) *J Ind Ecol 526-544.* www.sustainable-systems.org.uk

Materials in a vacuum cleaner

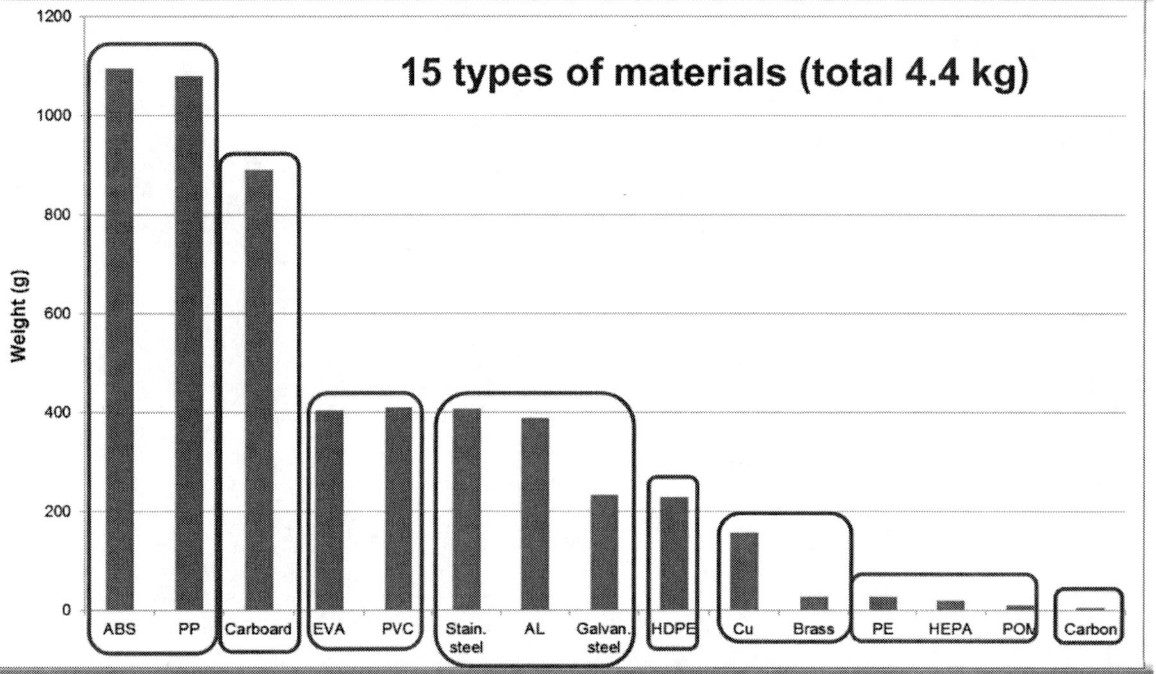

Gallego Schmid et al. (2016) *SciTot Env 559192-203.* www.sustainable-systems.org.uk

Waste management: Vacuum cleaners

- EU (small household appliances)
  - 37% collected as WEEE
  - 29% recycled/reused
- UK (small household appliances)
  - 21% collected as WEEE
  - 18% recycled/reused

Gallego Schmid et al. (2016) *SciTot Env 559192-203.* www.sustainable-systems.org.uk

Sustainable Industrial Systems

LCA of vacuum cleaners: Contribution of materials

Impact	Current electricity mix Lifetime: 8 years	2020 electricity mix Lifetime: 5 years
Carbon footprint	12%	25%
Depletion of metals and minerals	38%	43%
Depletion of fossil fuels	13%	27%

Gallego Schmid et al. (2016) *SciTot Env 559192-203.* www.sustainable-systems.org.uk

Sustainable Industrial Systems

Circularity indicators

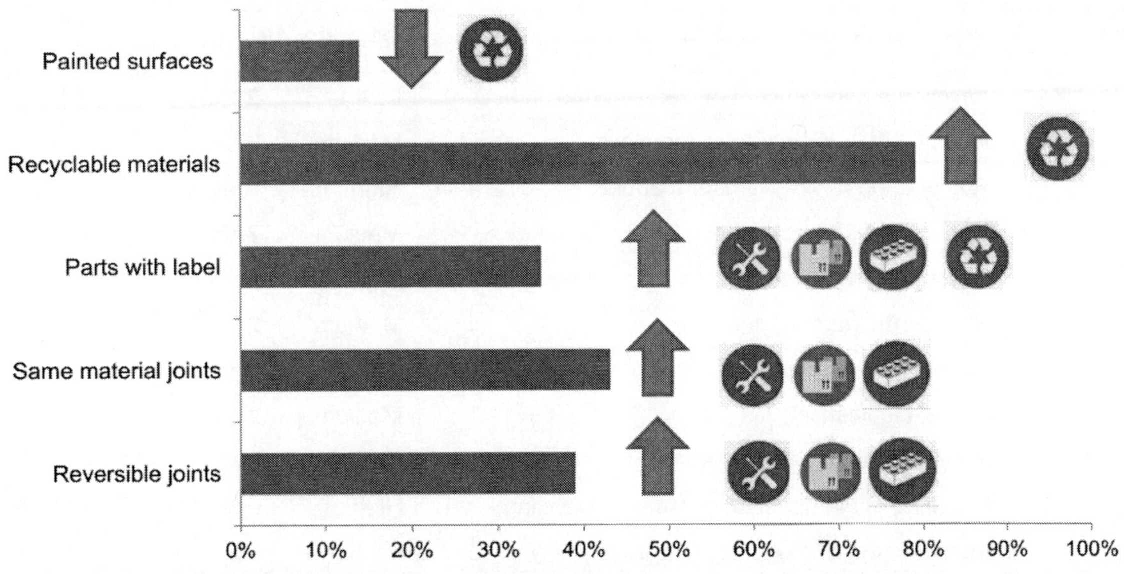

www.sustainable-systems.org.uk

Sustainable Industrial Systems

Improvement opportunities

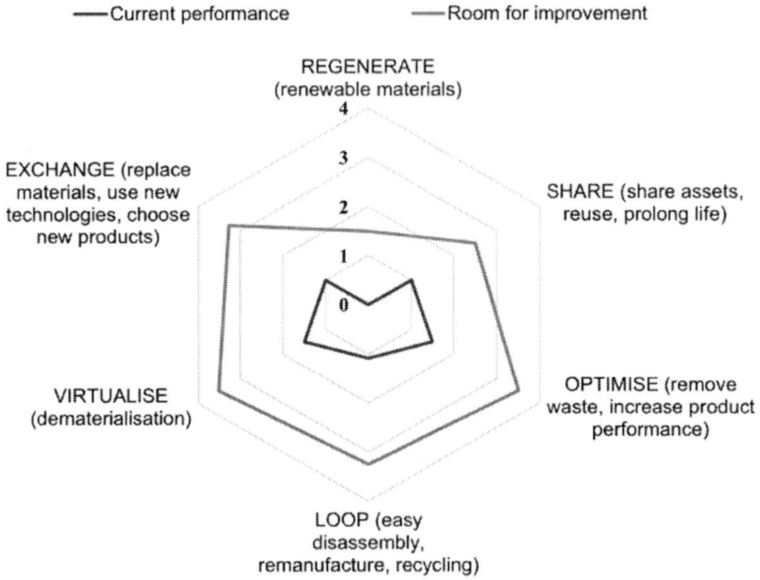

Mendoza et al. (2017) *J Ind Ecol 526-544.* www.sustainable-systems.org.uk

Sustainable Industrial Systems

Improvement opportunities

Actions	Potential circular economy solutions	Feasibility	Priority
VIRTUALISE	Envision a new concept of cleaning (multifunction)	Low	2025
	Remove the wiring system (cordless)	Medium	2020
SHARE	Repair shops; improve user manuals	High	Present
	Vacuum-for-life service (availability of spares)	Medium	2020
OPTIMISE	Buy according to your needs (e.g. accessories)	High	Present
	Reduce the number of elements	Medium	2020
	Standardise screws and pieces	High	Present
	Use of bioplastics	Medium	2025
	Implement a take-back system	Medium	2020
	Labelling of elements for reuse and recycling	High	Present
	Increase recycled content and recyclability	Medium	2020

Mendoza et al. (2017) *J Ind Ecol 526-544.* www.sustainable-systems.org.uk

Sustainable Industrial Systems

Strategy for implementing CE in practice: BECE

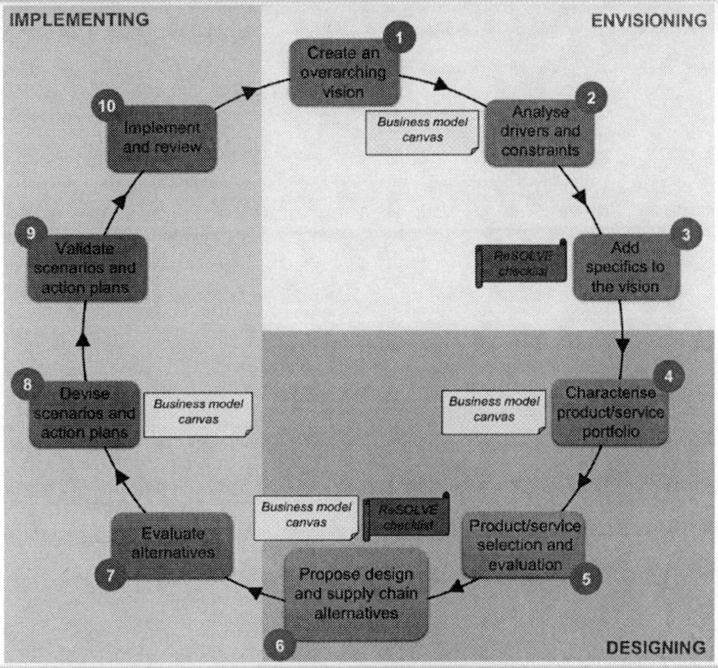

Mendoza et al. (2017) *J Ind Ecol 526-544.* www.sustainable-systems.org.uk

Sustainable Industrial Systems

Further details

METHODS, TOOLS, AND SOFTWARE

Integrating Backcasting and Eco-Design for the Circular Economy

The BECE Framework

Keywords:

Summary

Introduction

www.sustainable-systems.org.uk

Sustainable Industrial Systems

Conclusions

- Most products and services are not designed for CE
- Implementation of CE will be challenging but is achievable
  - Drivers are increasing
  - Methods and tools are available
  - A life cycle approach is essential
- More success stories are needed to stimulate the uptake
- Legislation will need to get tougher

www.sustainable-systems.org.uk

Life Cycle Assessment of calcium carbonate loop CO_2 capture technology for brown coal power plant unit of the Czech Republic

Authors: Kristína Zakuciová [a,b,*], Vladimír Kočí[a], Karel Ciahotný[a], Ana Carvalho[c]

[a] University of Chemistry and Technology, Prague, Czech Republic

[b] Research Centre Řež s.r.o, Czech Republic

[c] CEG-IST, Instituto Superior Técnico, Universidade de Lisboa, Portugal

Tel.: + 420 775 364 032

E-mail address: k.zakuciova@gmail.com

UNIVERSITY OF CHEMISTRY AND TECHNOLOGY PRAGUE

Centrum výzkumu Řež s.r.o. Research Centre Řež

TÉCNICO LISBOA

3.10.2017

Content

- Introduction and main goals of the work
- LCA methodology and case study description
- System boundaries and technical assumptions
- Considered scenarios
- Results
- Discussion and Conclusion

2

Introduction

- Europe goals for **low carbon economy** ⇒ SET plan with 7 roadmaps

⇓

- **United Nations Framework Convention on CC** ⇒ **Carbon Capture and Storage (CCS) Technologies**

Life Cycle Assessment (LCA)- effective tool for understanding environmental impacts and optimization of processes

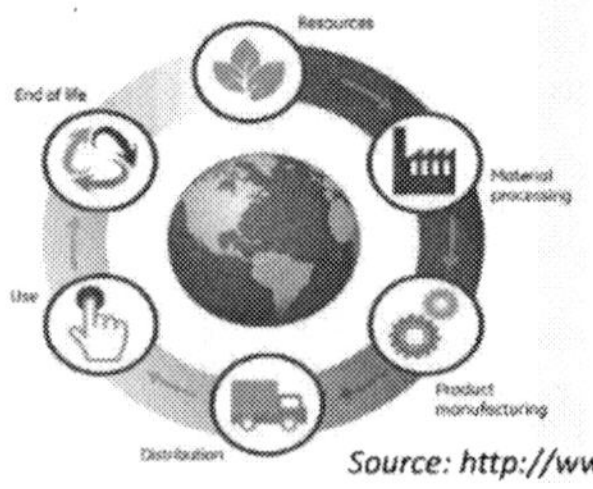

Source: http://www.project-sas.com

3

Goals of study

- ➢Assessment of various scenarios for calcium carbonate CCT for modelled 250MWe power unit.
- ➢Identifying major processes with the highest environmental contribution.
- ➢Single Score Analysis - evaluation of sum of impacts in different LCA methods (CML, ReCiPe, IMPACT, EDIP).
- ➢Suggestions for optimization.

4

LCA methodology – ISO 14044

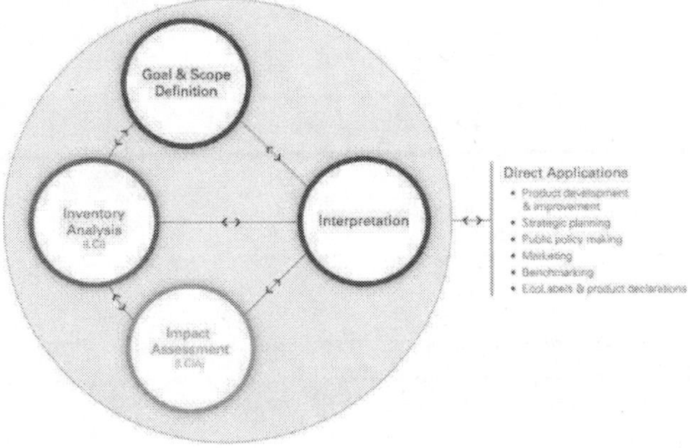

Source:
http://www.lcanz.org.nz/introduction-lca

1. Goal and scope: EI of capturing CO_2 by calcium looping process in a connection with an existent Czech power unit, of gross capacity 250 MWe, which is based on brown coal as source of energy.

- Functional unit: the nominal power output, 250MWe/year
- Definition of boundaries – PP and CCS chain, ***excluding*** transport and storage of CO_2

2. Inventory analysis: Data for the Czech power unit was collected from its annual operations' report
Data for carbonate loop - calculated according to the taken assumptions

3. LCIA: CML and ReCiPe method characterization, normalization , Pareto analyses and Single Score Analysis.

5

Case study- CO_2 reduction in the Czech Republic

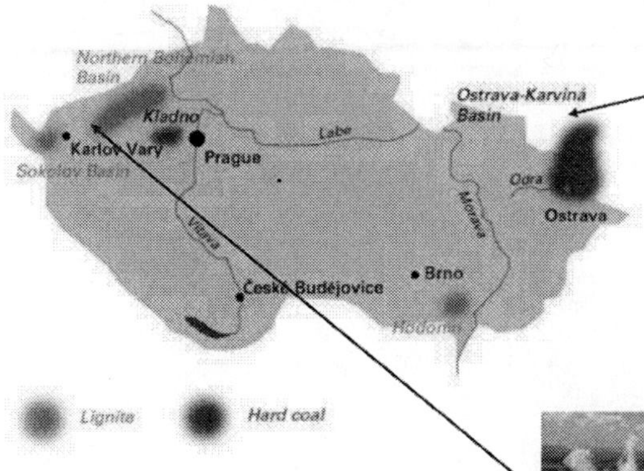

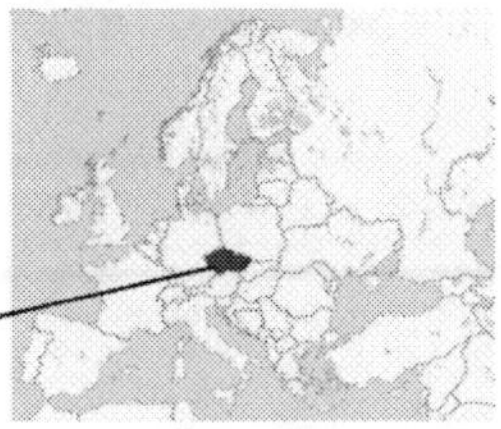

In 2015, approximately 51.4% of total gross electricity production (83.9 TWh) was generated from coal, 32.0% from nuclear energy and 7.0% from solar, hydro and wind

PP Prunerov: Rebuilt of three 250MWe power unit; increase of energy efficiency to 40 %. Innovation seen in CCT implementation.

6

System boundaries and assumptions

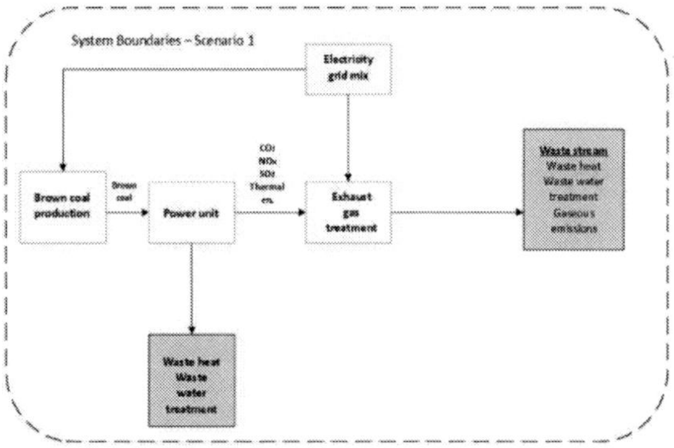

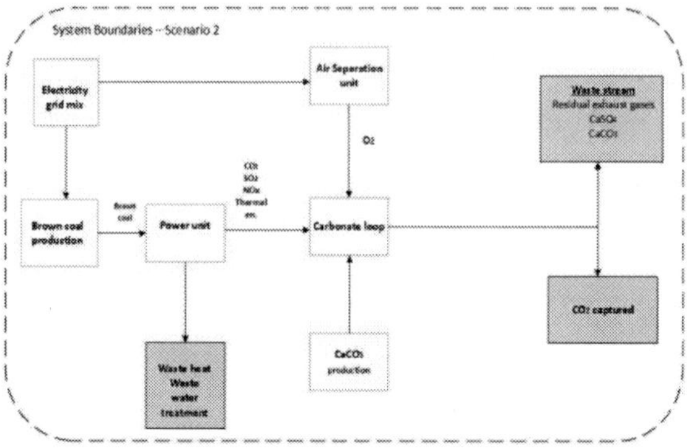

Assumptions for CaO looping system:

- The carbon capture is assumed to be 90 % CO_2 from flue gases.
- $CaSO_4$ is considered as inert waste in landfill.
- Thermal energy from carbonation process in scenario 2.1 and 2.2 is in 100% considered as a steam and represents credit in for environmental profile.
- 85 % of produced steam in scenario 2.3 is used as return heat for calcium carbonate.
- Distance between the power plant and landfill is assumed to be 50 km.
- Distance between the place of brown coal mining and processing to the power plant is assumed to be 100 km.
- Distance between the place of $CaCO_3$ mining and processing to the power plant is assumed to be 120 km.

7

Carbonate loop capture system

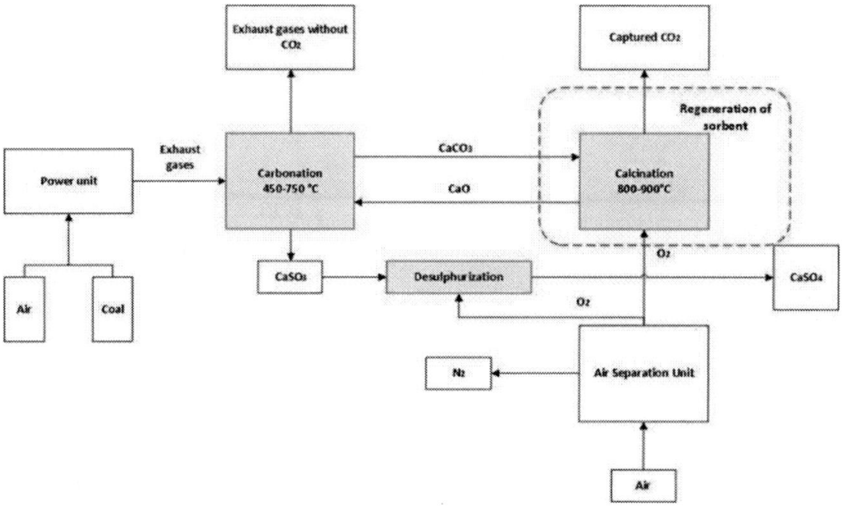

8

Considered scenarios for carbonate loop system

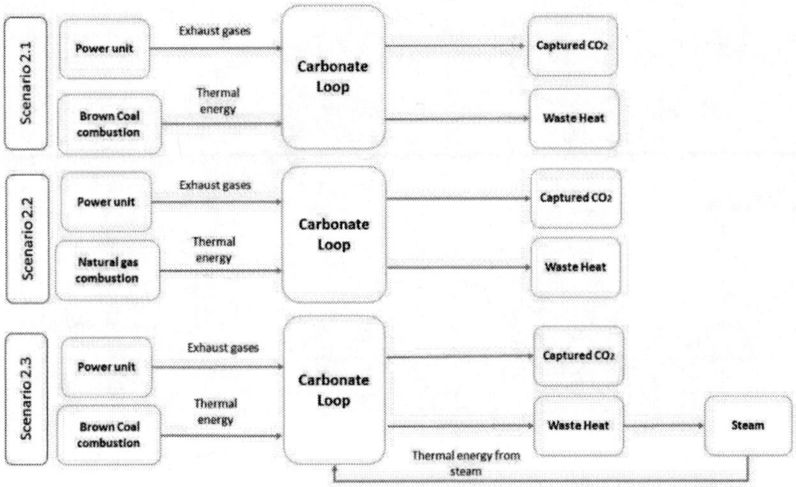

9

Results – Scenario 1

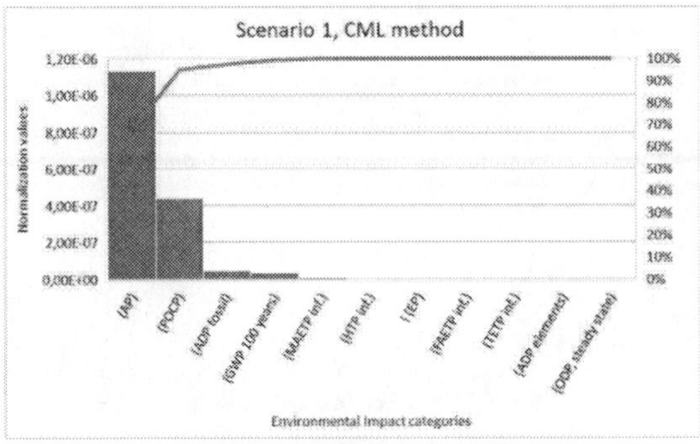

	Brown Coal production	Diesel mix at refinery	Municipal waste water treatment	SO2 emissions
AP	7,78E-10	8,97E-11	4,3E-12	1,12484E-06
POCP	7,16E-10	1,27E-10	2,23E-12	4,36931E-07

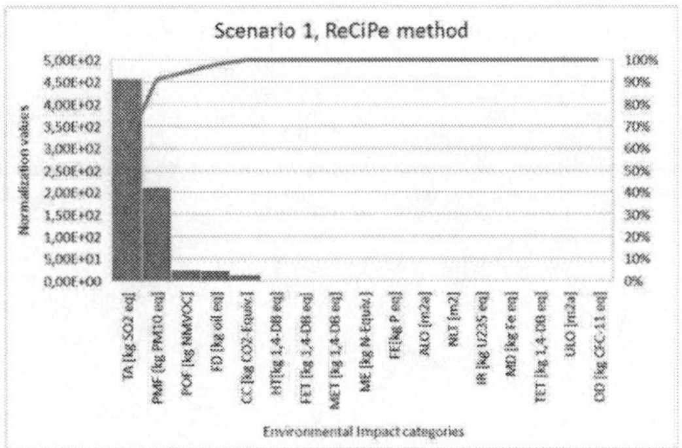

	Brown Coal Production	Diesel mix at refinery	Municipal waste water treatment	SO2 emissions
TA	0,234	0,0263	0,00202	458
PMF	0,339	0,0419	0,00138	211

10

Results – Scenario 2.1

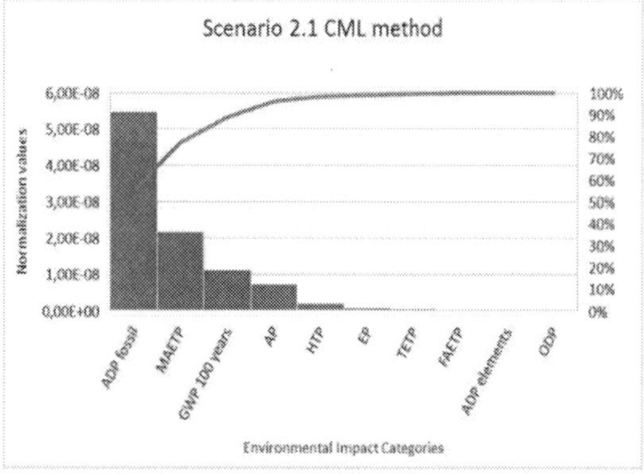

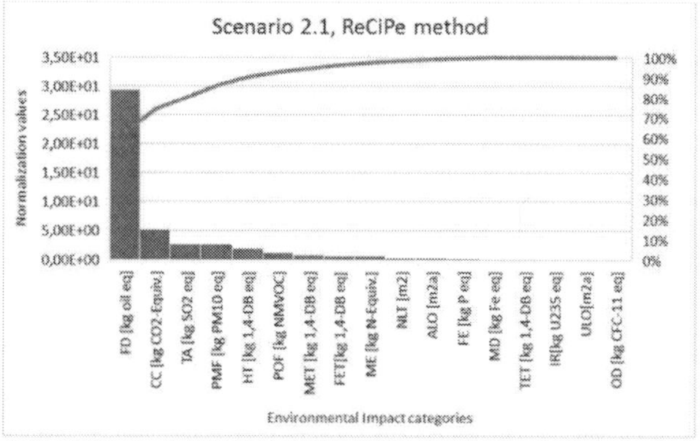

11

Results – Scenario 2.1

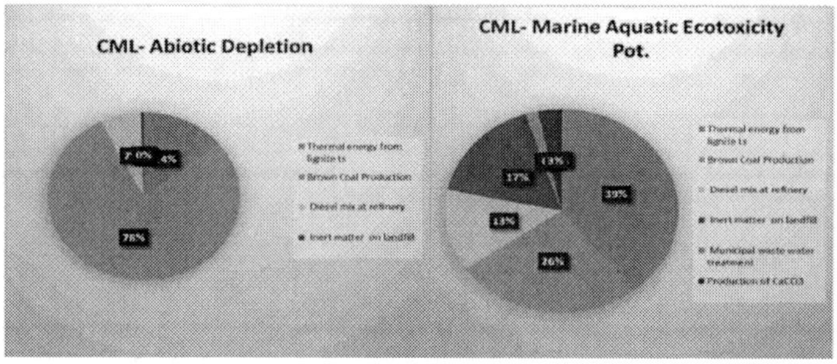

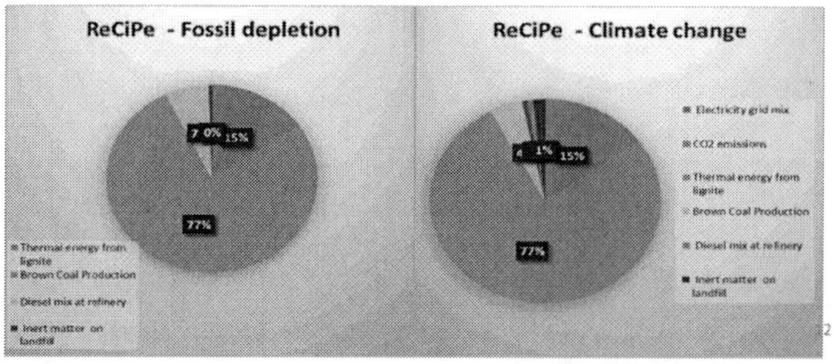

22

Results – Scenario 2.2

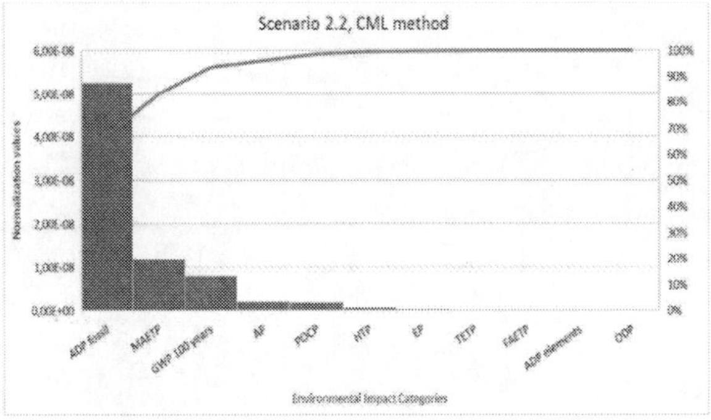

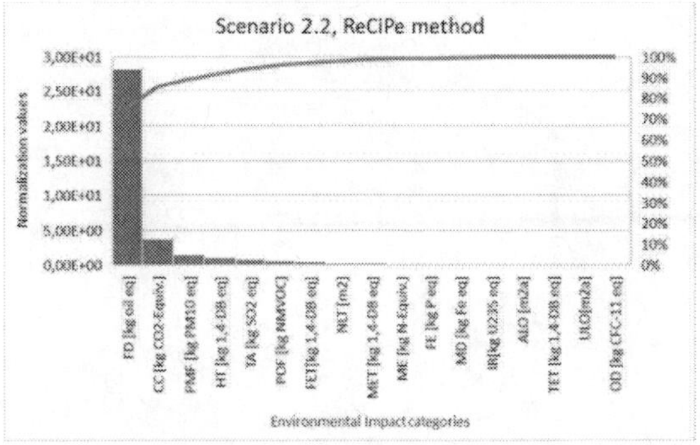

13

Results – Scenario 2.2

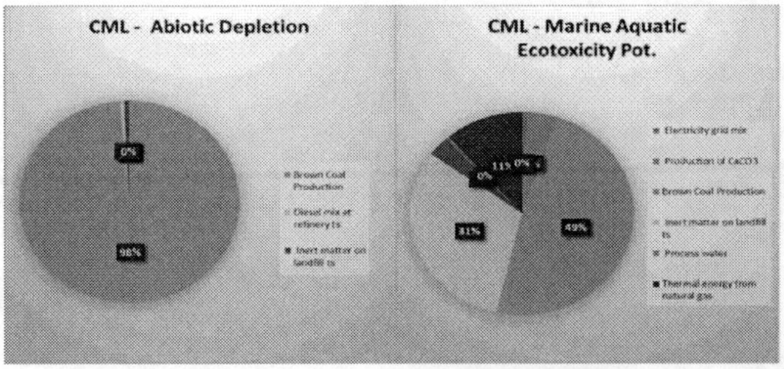

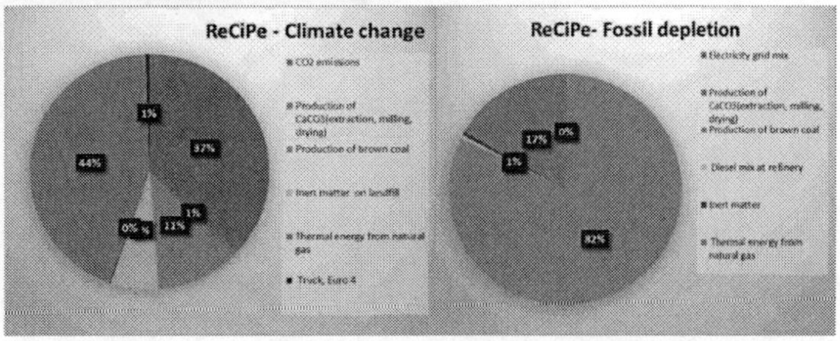

14

Results – Scenario 2.3

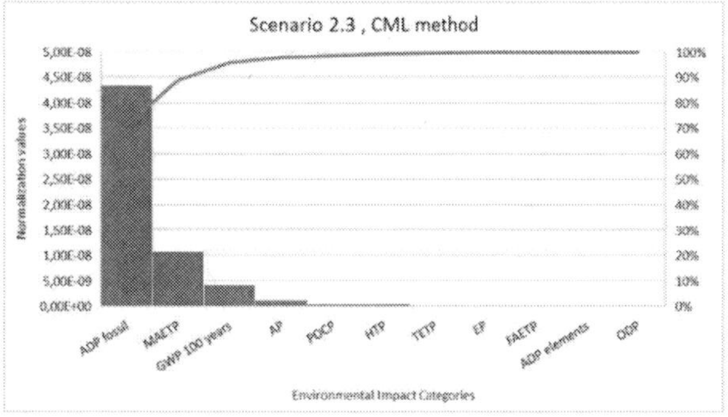

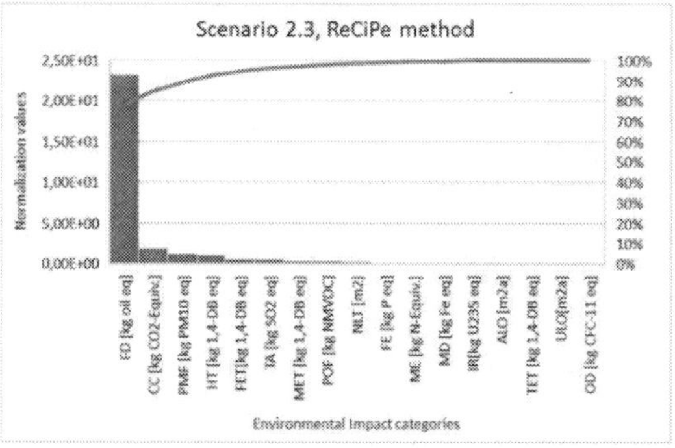

15

Results - Scenario 2.3

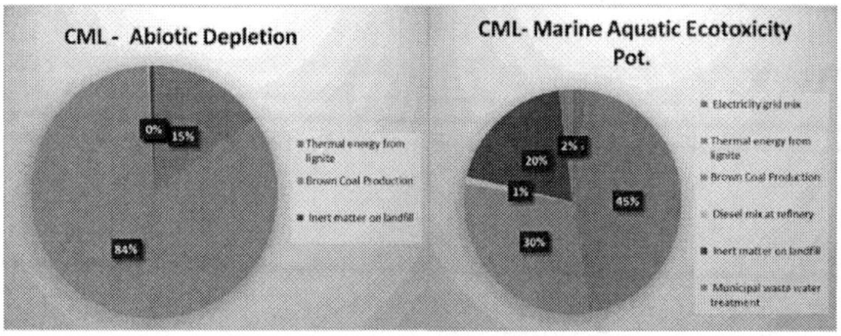

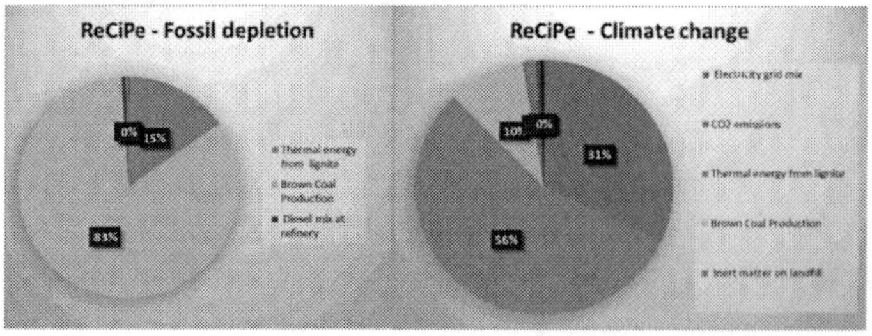

16

Single Score Analyses

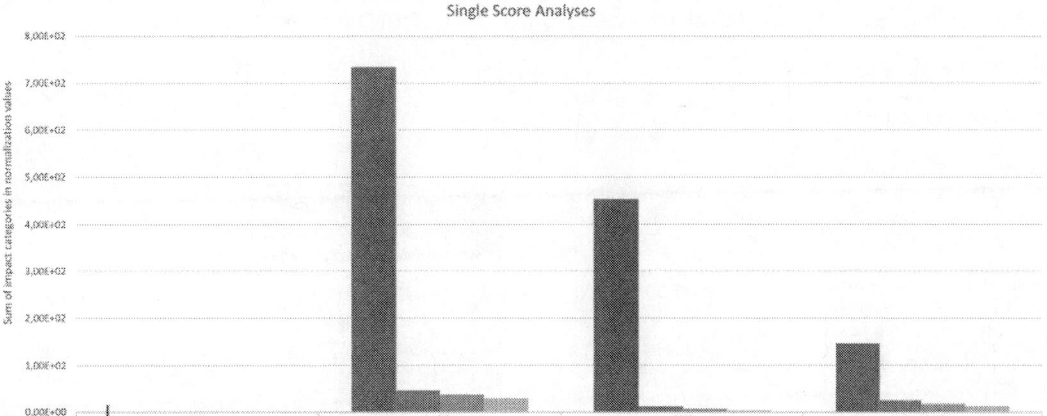

	CML	ReCiPe	EDIP	Impact 2002+
Scenario 1	1,65E-06	7,34E+02	4,54E+02	1,47E+02
Scenario 2.1	9,59E-08	4,65E+01	1,20E+01	2,52E+01
Scenario 2.2	7,74E-08	3,73E+01	7,03E+00	1,73E+01
Scenario 2.3	6,09E-08	2,97E+01	3,73E+00	1,27E+01

17

Discussion and conclusions

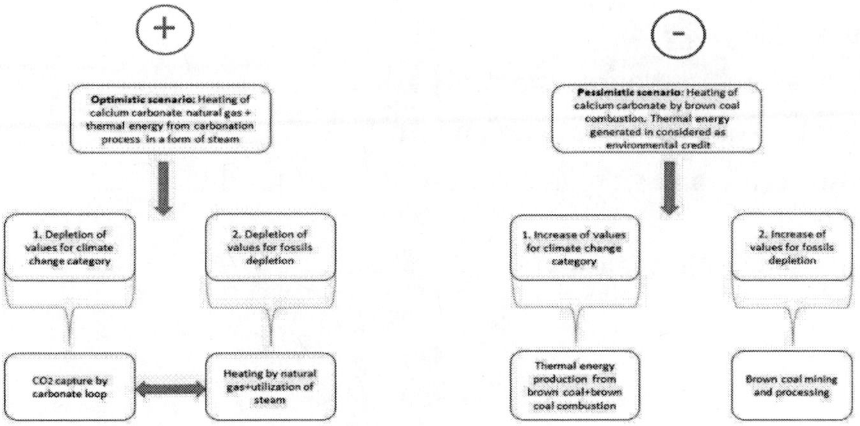

18

Scenario 1 vs Scenario 2

- Fossils depletion (FD) lower in Scenario 1 in compare with sc. 2.1, 2.2
- FD lowest in Scenario 2.3
- **Scenario 2** results in:

Depletion in global warming potential (kg CO_2-equiv.)	Depletion of acidiphication potential (kg SO_2-equiv.)	Depletion of photochemical ozone formation(kg ethene-equiv.)
• Capture of CO_2 and heating by natural gas plus + steam utilization	• Carbonate loop as desulphurization unit	• Lowering level of mining intensification by utilization of natural gas

With or without CCS?

- Innovative technology **BUT**:
- Need for effective energy source (natural gas)
- Effective utilization of waste heat (steam form)
- Further infrastructure and storage solutions for captured CO_2 transportation
- Robust socio – economic analyses

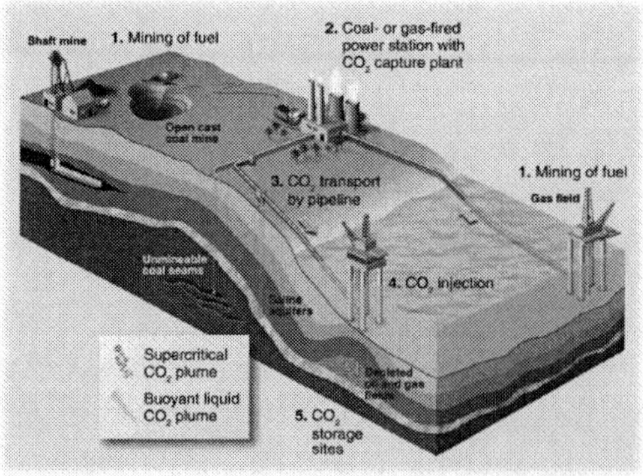

Source: http://www.ukccsc.co.uk

Thank you for your attention

k.zakuciova@gmail.com

21

Handling multi-scale/multipartner information for biorefinery modelling, integration and sustainability assessment

Mathias Janssen[a] , Elin Svensson[b], Stavros Papadokonstantakis[c]

[a]Department of Technology, Management and Economics
[b]CIT Industrial Energy
[c]Department of Space, Earth, and Environment

Chalmers University of Technology
Gothenburg, Sweden

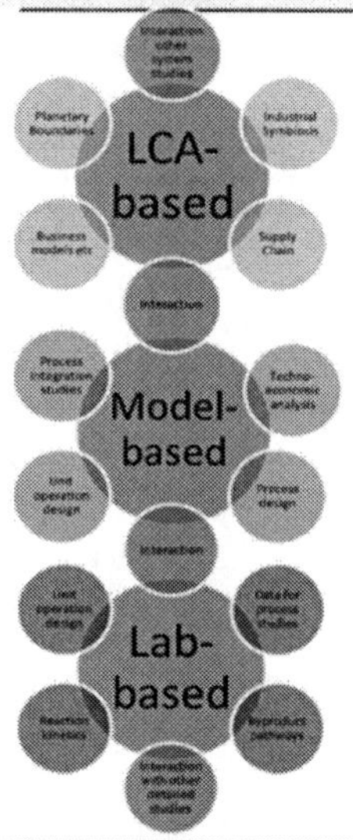

Main interaction focus:
- Introducing planetary boundaries and industrial symbiosis in the LCA framework
- Using LCA as optimization tool (feedback loops to model-based process design)

Main interaction focus:
- Extending process integration and intensification from technoeconomic analysis to LCA (feedback loops to LCA-based design)
- Using plant oriented process models as experimentation guide (feedback loops to lab-based design)

Main interaction focus:
- Creating targeted experimentation protocols for the generation of relevant process and equipment design data
- Connecting lab-based design with theoretical frameworks (e.g. CAMD, reaction networks)

- Typically different research groups with different expertise, focus and goals at different scales or knowledge domains
- Typically groups need to collaborate in small to large size transdisciplinary (or interdisciplinary) projects for a comprehensive research output
- The middle layer of (process) modelling is of high importance for an effective communication between scales/partners (although sometimes neglected)
- Research on the potential of biorefineries is a good example

BIOBUF

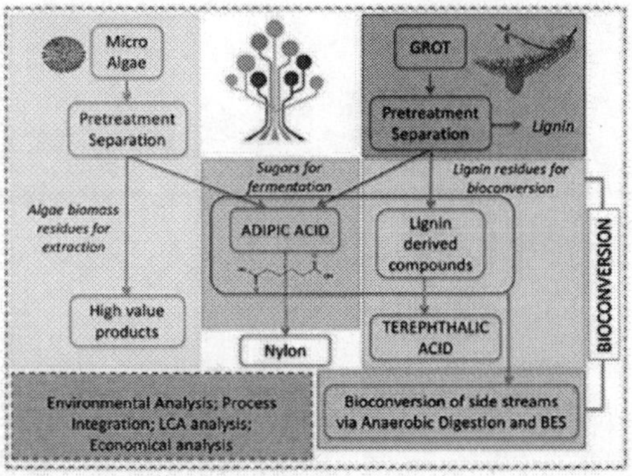

Upgrading of renewable domestic raw materials to value-added bulk and fine chemicals for a biobased economy: technology development, systems integration and environmental impact assessment.

- **Bio**refinery for **Bu**lk and **F**ine Chemicals
- Involves collaboration of:
  Industrial/Algal Biotechnology
  (BBE dept. – Micro/Lab scale)
  Industrial/Energy Technology
  (SEE dept./CIT – Meso/Process Scale)
  Environmental Systems Analysis
  (TME dept./RISE – Macro/LCA Scale)

Level of Detail / TRL

BIOBUF

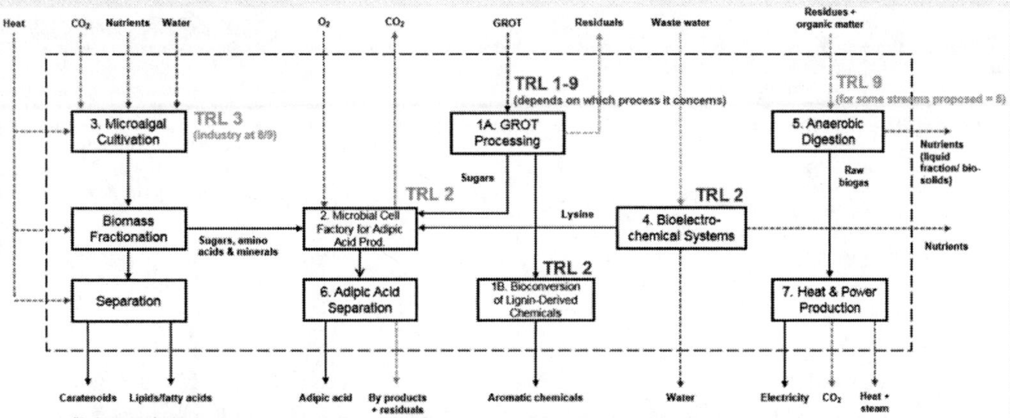

Project members:

1. Cecilia Matterson & Hans Theliander + Salvatore Fusco & Calle Franzén
2. Emma Karlsson & Valeria Mapelli
3. Joshua Mayers & Eva Albers
4. Nikolaos Xafenias & Valeria Mapelli
5. Christel Kampman & Maurizio Bettiga
6. Stavros Papadonkonstantakis & Mathias Janssen
7. Mathias Janssen

LCA: Anna Ekman: Algal processes & Christel Cederberg

CHALMERS UNIVERSITY OF TECHNOLOGY

Conventional ADIPIC ACID Route

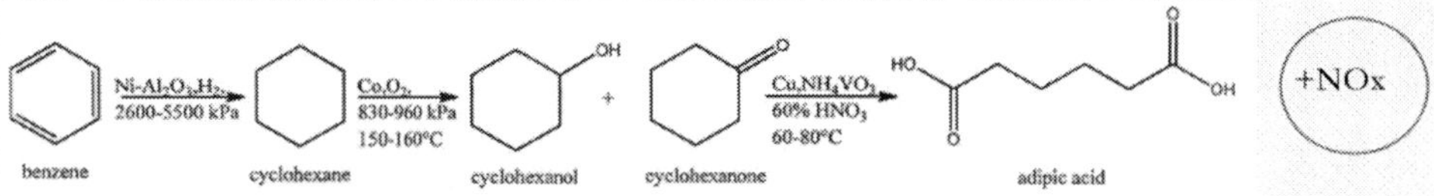

- One applied large-scale chemical pathway. [1]

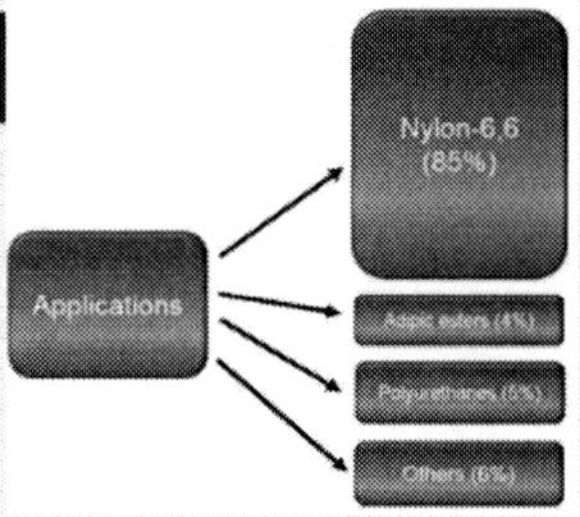

	Adipic Acid	Lactic Acid	Succinic Acid
Current production	2 600'000 t/year	300'000 t/year	25'000 t/year
Potential growth	3-5 %/year	19 %/year	10 %/year
Current market prize	1600 $/t	1500 $/t	2500 $/t

Companies such as Verdezyne, DSM, BioAmeber, Ameris, Aemethis, Genomatica and Rennovia have designed production technologies to develop 100% bio AA over the coming years.

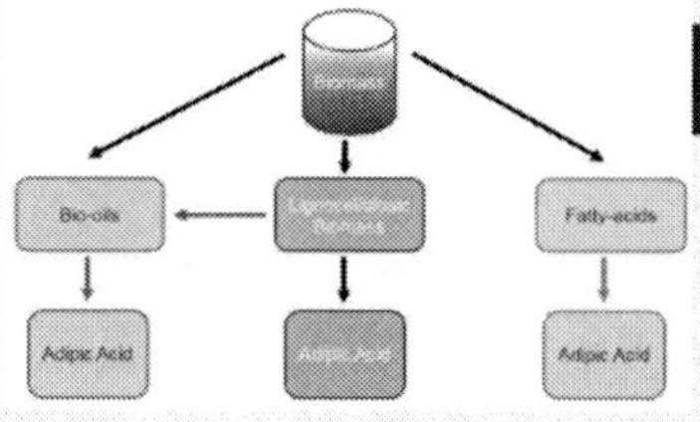

[1] T. Polen, M. Spelberg, M. Bott, 2013, Toward biotechnological production of adipic acid and precursors from biorenewables, Journal of Biotechnology

CHALMERS UNIVERSITY OF TECHNOLOGY

ICOSSE'17 / WCCE 2017, October 3rd

Biotechnological ADIPIC ACID Route

BIOBUF

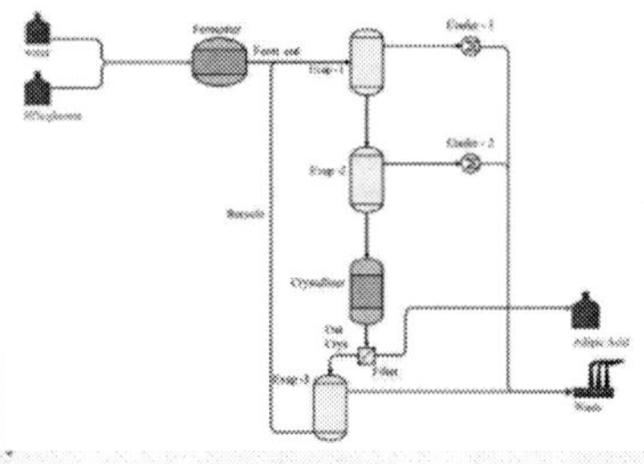

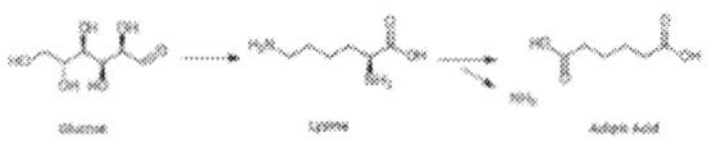

	Scenario 1	Scenario 2
Water	40 mM	400 mM
Conversion	15,4 wt%	20,8 wt%

Component	Yield
Glucose	0.426
Lysine	0.028
Dihydroxyacetone	0.026
Glycerol	0.004
Trehalose	0.003
Acetate	0.045
Glycine	0.007
NH_3	0.253
Adipic Acid	0.253

Preliminary Environmental/Economic Assessment

OPEX: 3.6-6.8 $/kg-AA
CAPEX not included

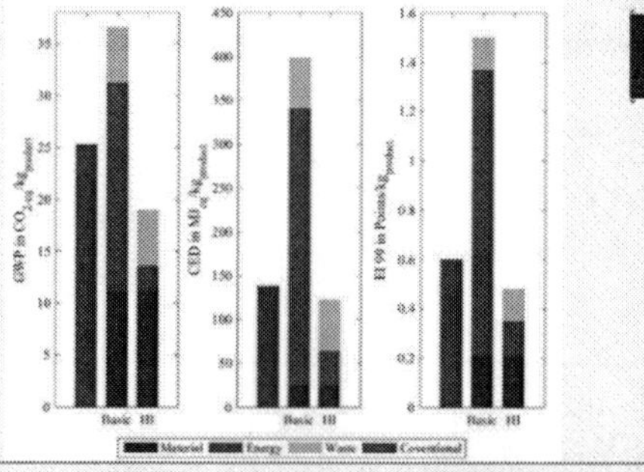

Solubility of Adipic Acid
(detailed data available)

$$C_6H_{10}O_4(aq) + H_2O \rightleftharpoons C_6H_9O_4^- + H_3O^+$$

$$C_6H_{10}O_4(s) \rightleftharpoons C_6H_9O_4^- + H^+$$

$$\ln(K_{eq}) = A + B/T + C/ln(T) + D \cdot T$$

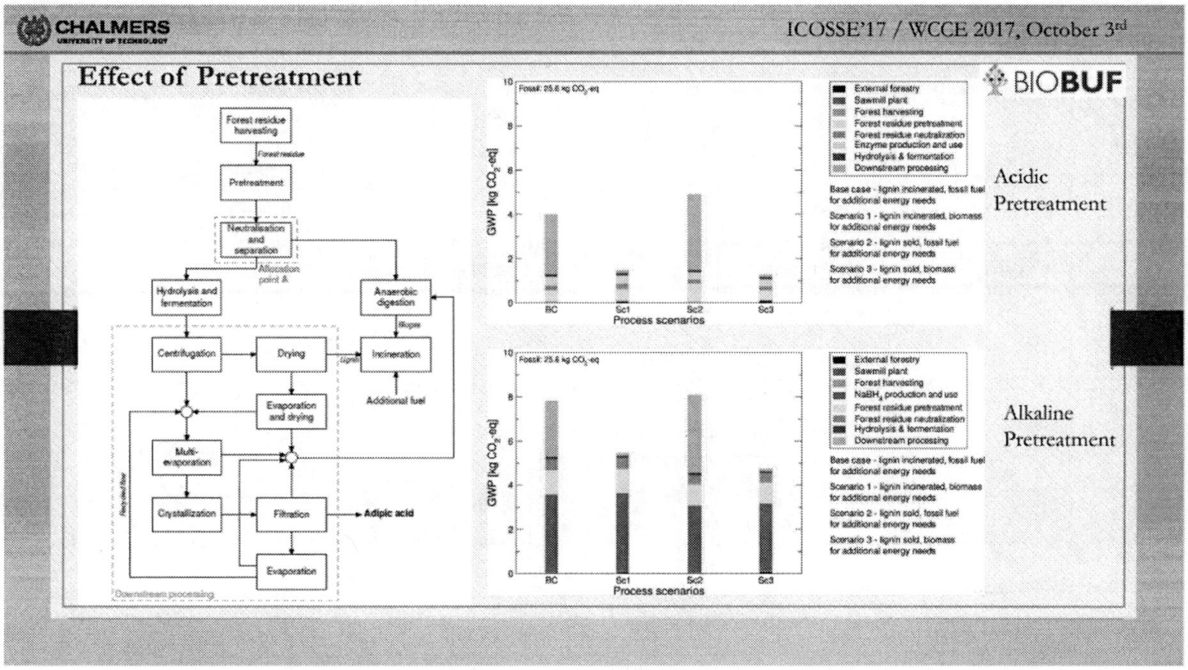
CHALMERS
UNIVERSITY OF TECHNOLOGY
ICOSSE'17 / WCCE 2017, October 3rd
Effect of Pretreatment
BIOBUF
Forest residue harvesting
Pretreatment
Neutralisation and separation
Hydrolysis and fermentation
Anaerobic digestion
Centrifugation
Drying
Incineration
Additional fuel
Evaporation and drying
Multi-evaporation
Crystallization
Filtration
Adipic acid
Evaporation
Downstream processing
GWP [kg CO2-eq]
Process scenarios
BC
Sc1
Sc2
Sc3
External forestry
Sawmill plant
Forest harvesting
Forest residue pretreatment
Forest residue neutralization
Enzyme production and use
Hydrolysis & fermentation
Downstream processing
Base case - lignin incinerated, fossil fuel for additional energy needs
Scenario 1 - lignin incinerated, biomass for additional energy needs
Scenario 2 - lignin sold, fossil fuel for additional energy needs
Scenario 3 - lignin sold, biomass for additional energy needs
Acidic Pretreatment
Alkaline Pretreatment

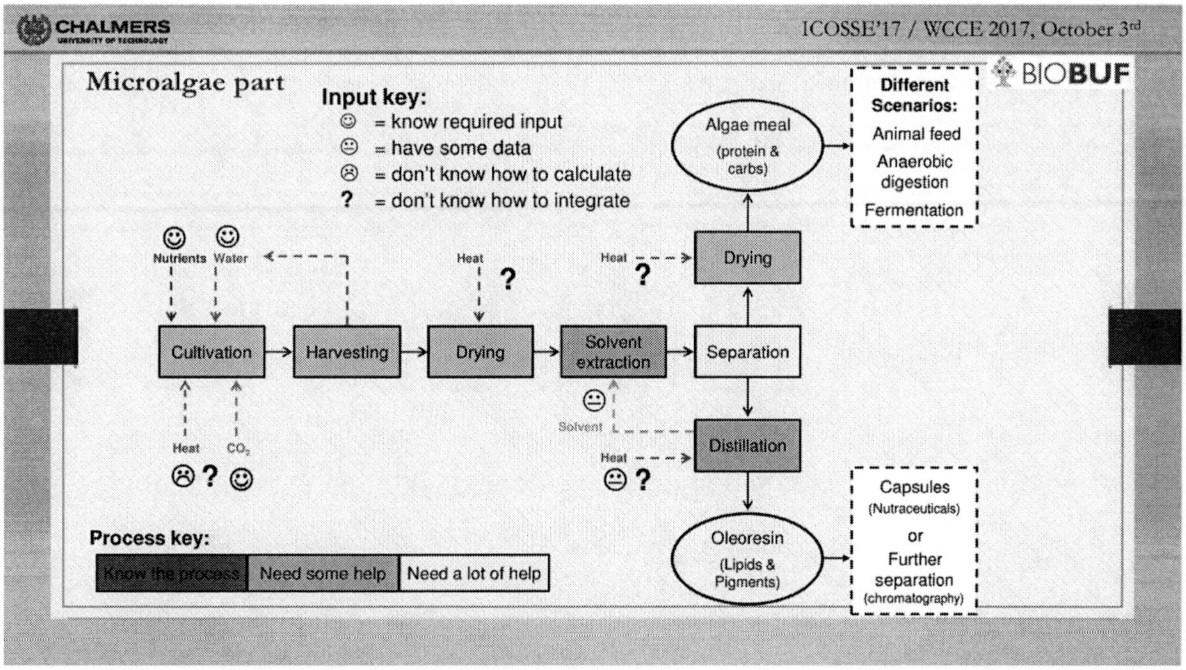
CHALMERS
UNIVERSITY OF TECHNOLOGY
ICOSSE'17 / WCCE 2017, October 3rd
Microalgae part
BIOBUF
Input key:
☺ = know required input
😐 = have some data
☹ = don't know how to calculate
? = don't know how to integrate
Nutrients
Water
Heat
Cultivation
Harvesting
Drying
Solvent extraction
Separation
Drying
Distillation
Solvent
CO2
Algae meal (protein & carbs)
Different Scenarios:
Animal feed
Anaerobic digestion
Fermentation
Oleoresin (Lipids & Pigments)
Capsules (Nutraceuticals)
or
Further separation (chromatography)
Process key:
Know the process
Need some help
Need a lot of help

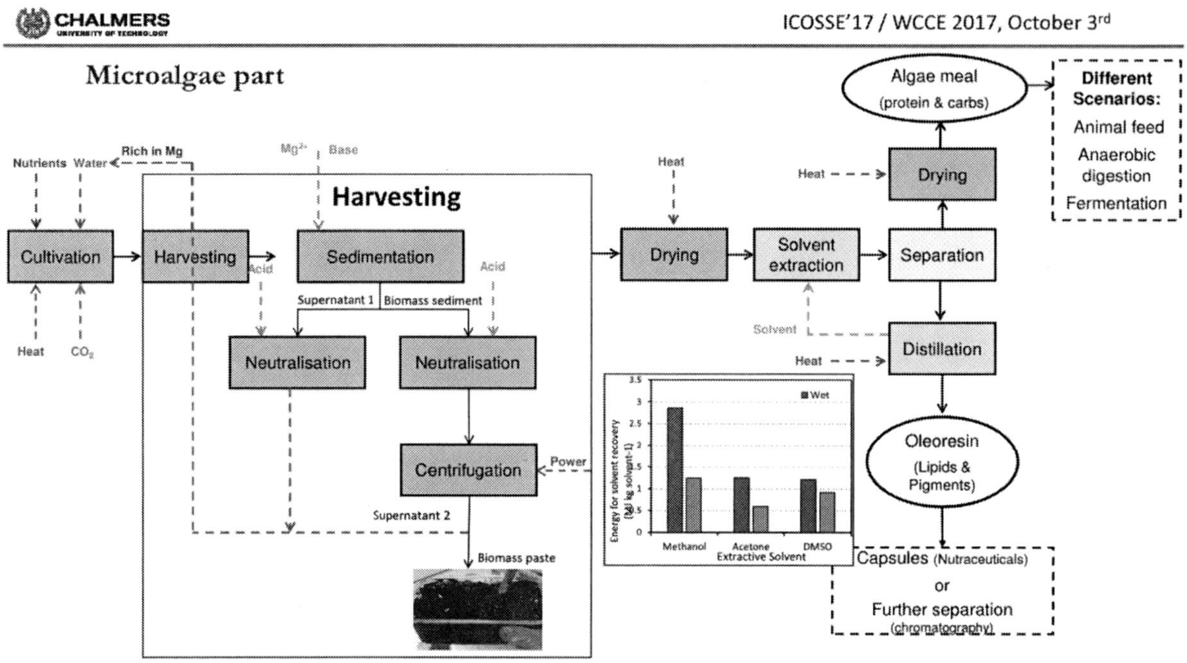
CHALMERS
ICOSSE'17 / WCCE 2017, October 3rd
Microalgae part
Nutrients
Water
Rich in Mg
Cultivation
Heat
CO2
Harvesting
Mg2+
Base
Sedimentation
Acid
Acid
Supernatant 1
Biomass sediment
Neutralisation
Neutralisation
Centrifugation
Power
Supernatant 2
Biomass paste
Heat
Drying
Solvent extraction
Solvent
Separation
Heat
Drying
Algae meal (protein & carbs)
Different Scenarios:
Animal feed
Anaerobic digestion
Fermentation
Heat
Distillation
Oleoresin (Lipids & Pigments)
Capsules (Nutraceuticals)
or
Further separation (chromatography)
Wet
Methanol
Acetone
DMSO
Extractive Solvent

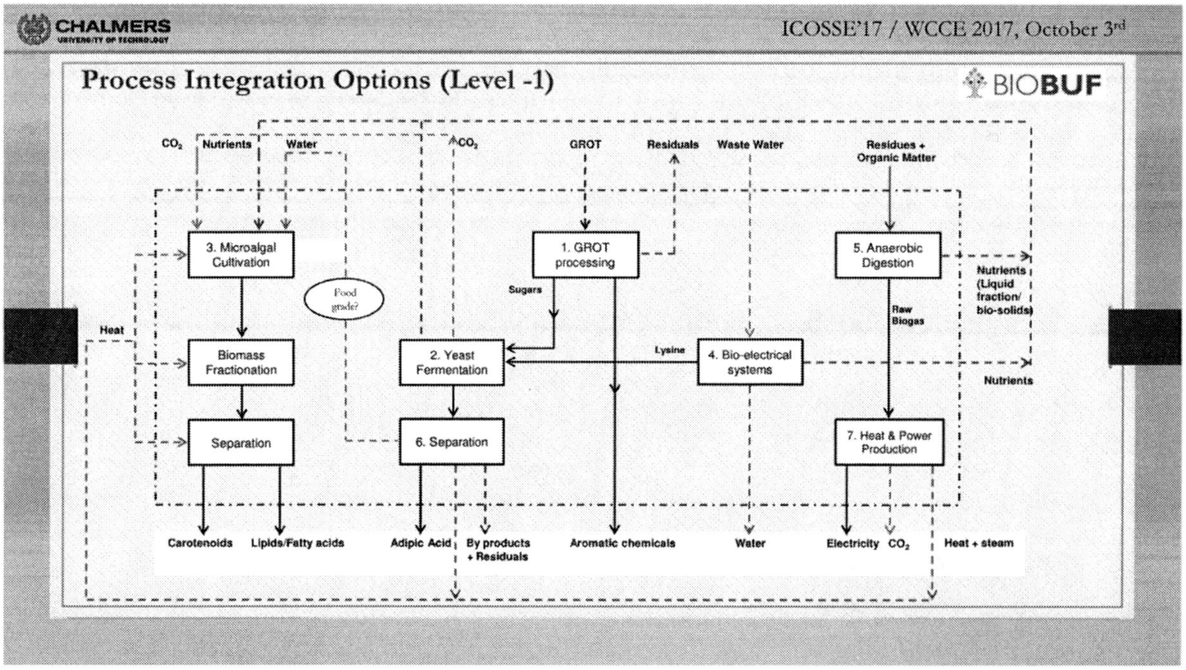
CHALMERS
ICOSSE'17 / WCCE 2017, October 3rd
Process Integration Options (Level -1)
BIOBUF
CO2
Nutrients
Water
CO2
GROT
Residuals
Waste Water
Residues + Organic Matter
3. Microalgal Cultivation
1. GROT processing
5. Anaerobic Digestion
Food grade?
Sugars
Raw Biogas
Nutrients (Liquid fraction/ bio-solids)
Heat
Biomass Fractionation
2. Yeast Fermentation
Lysine
4. Bio-electrical systems
Nutrients
Separation
6. Separation
7. Heat & Power Production
Carotenoids
Lipids/Fatty acids
Adipic Acid
By products + Residuals
Aromatic chemicals
Water
Electricity
CO2
Heat + steam

Process Integration Options (Level-2)

BIOBUF

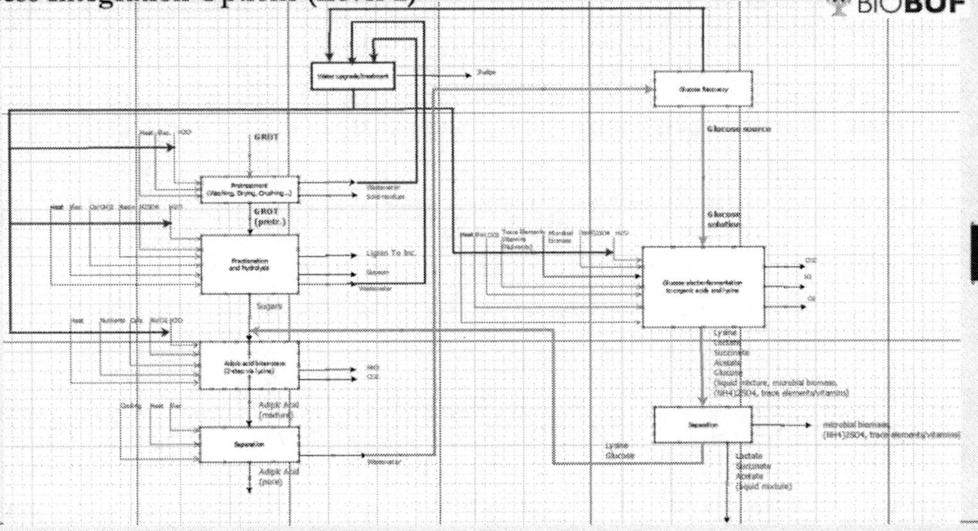

Conclusions

- Biorefinery design is not just about integrating (looping) processes (material/heat source to sink approaches), but also about **integrating information from/to different prividers to/from different users.**

- While **systematic methodologies** for complex material/energy recycling techniques exist, the same is **not true for information exchange** between lab/process/sustainability assessment (often very case specific).

- It is likely that this procedure is **best cordinated by the meso-scale process modelling and design**, involving multiple iterations between partners in the project life-time.

Algae as an Animal Feed Supplement: Nutrient Recovery and Water Reuse

Sage R. Hiibel
Chemical and Materials Engineering
University of Nevada, Reno
OC-88881

Antonia P. Faciola
Animal Science
University of Florida

M. Toufiq Reza
Mechanical Eng.
Ohio University

Charles J. Coronella
Chemical and Materials Eng.
University of Nevada, Reno

10th World Congress of Chemical Engineering
International Congress on Sustainability Science and Engineering
Barcelona, Spain ~ October 3, 2017

Where is UNR?

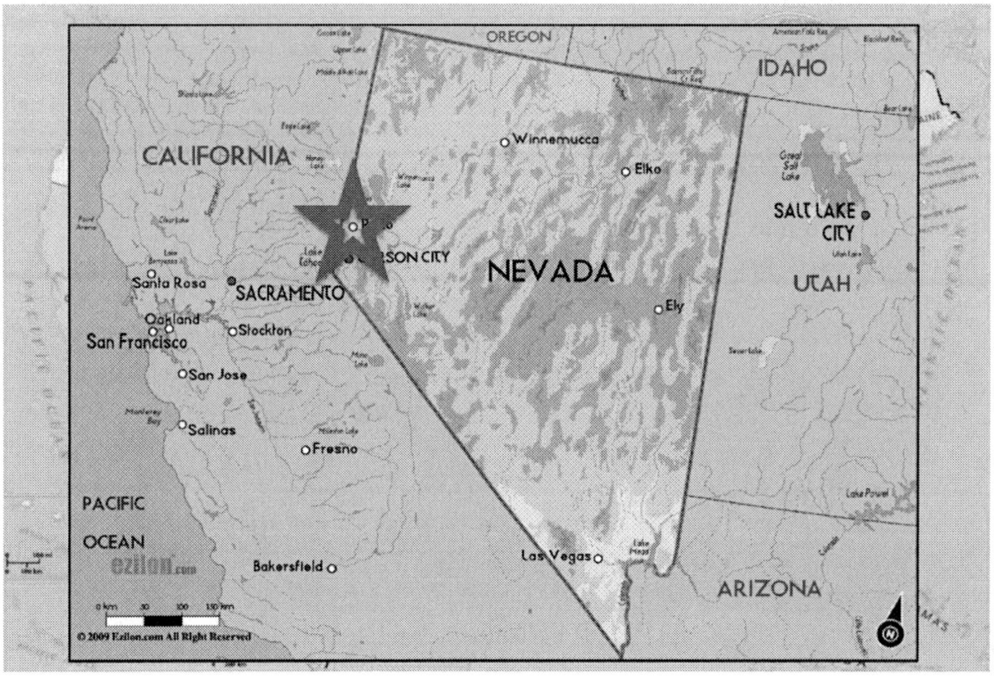

Project Motivation

- Majority of animal feed derived from terrestrial crops
  - Large physical foot print
    - Competition for other land use
    - Loss of natural habitat
  - High nutrient requirements
    - Limited phosphorus supply
    - Anthropogenic eutrophication

Sustainable alternative needed!

Algae as an Alternative?

- Non-coastal application → microalgae
  - Single-celled photosynthetic organisms
  - High protein content
  - Recognized human health supplement
- Smaller physi
  - Can utilize n
  - Opportunity ltivation
- High nutrient requirements
- Large water d

Addressing the Challenges

- Alternative nutrient sources

- Alternative water supplies

- High nutrient requirements
- Large water demand

Scott's Organic Fertilizer

12.5 vol% centrate from local WWTP

Day 1 | Day 10 | Day 17

Neochloris oleoabundans in 2,000 L indoor raceway ponds

Hiibel, et al. (2015) Frontiers in Energy Research

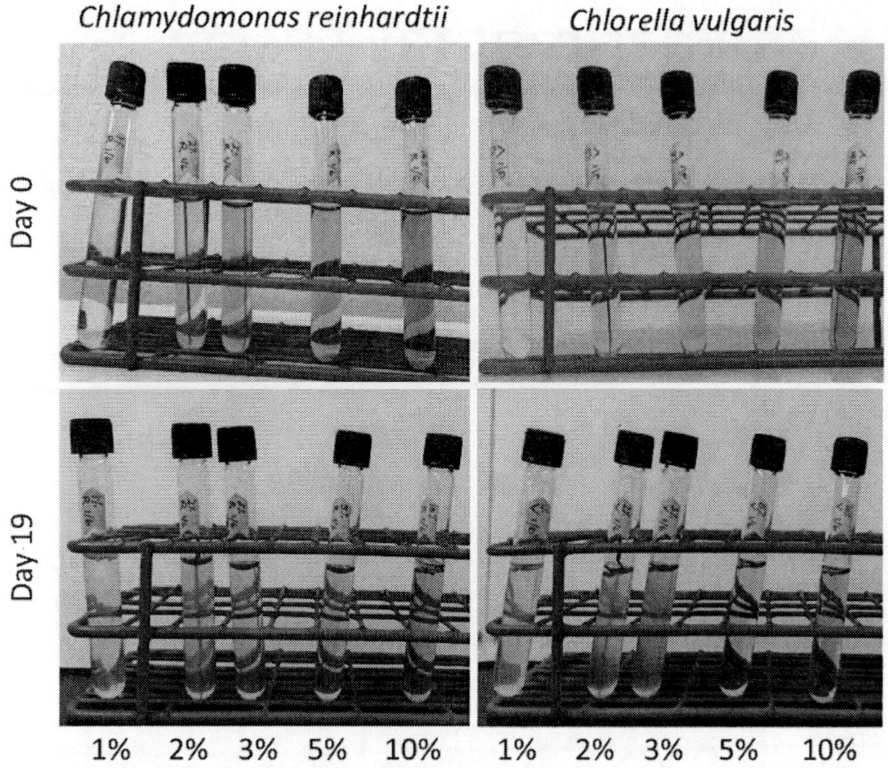

Aqueous by-product from hydrothermal carbonization of dairy manure

See poster by CJ Coronella for more details

Addressing the Challenges

- High nutrient requirements
  - Alternative nutrient sources
  - Nutrient recovery
    - Benefits of feed vs. fuel end use
- Large water demand
  - Alternative water supplies
  - Water reuse post harvest
    - Treatment requirements?

Experimental Design

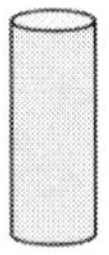

Experimental Design

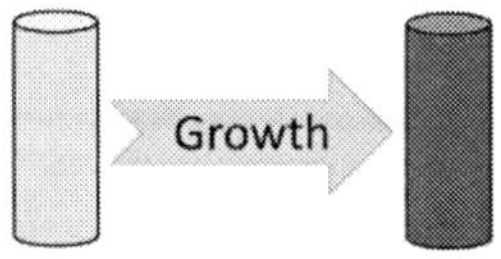

18:6 hr light:dark
21 °C
Constant agitation
No aeration or CO_2

Experimental Design

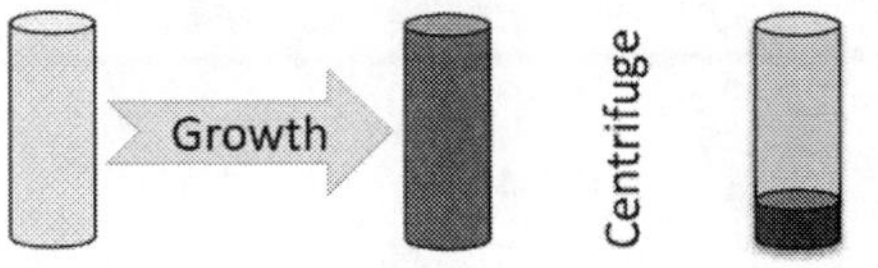

Experimental Design

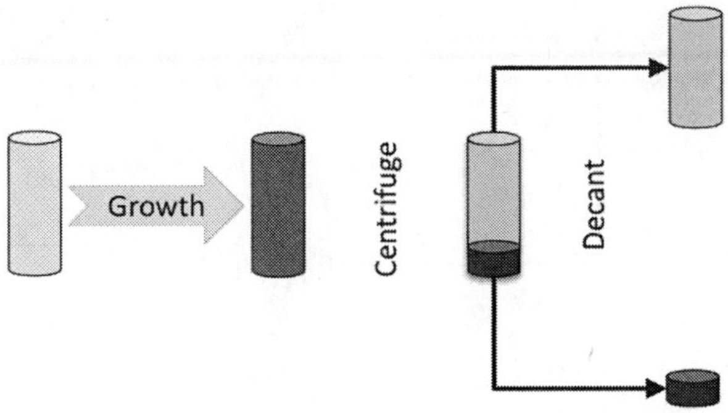

Experimental Design

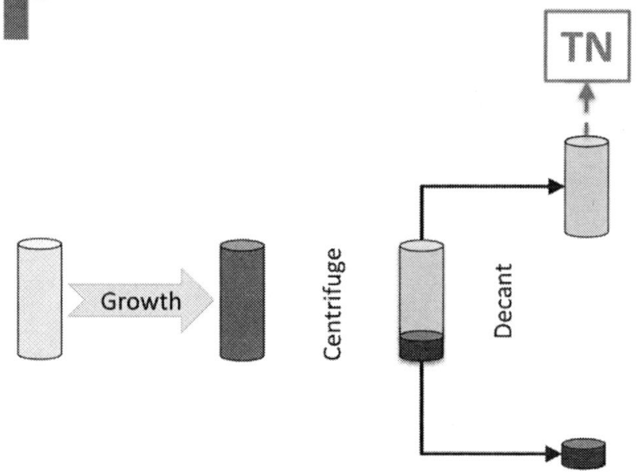

Experimental Design

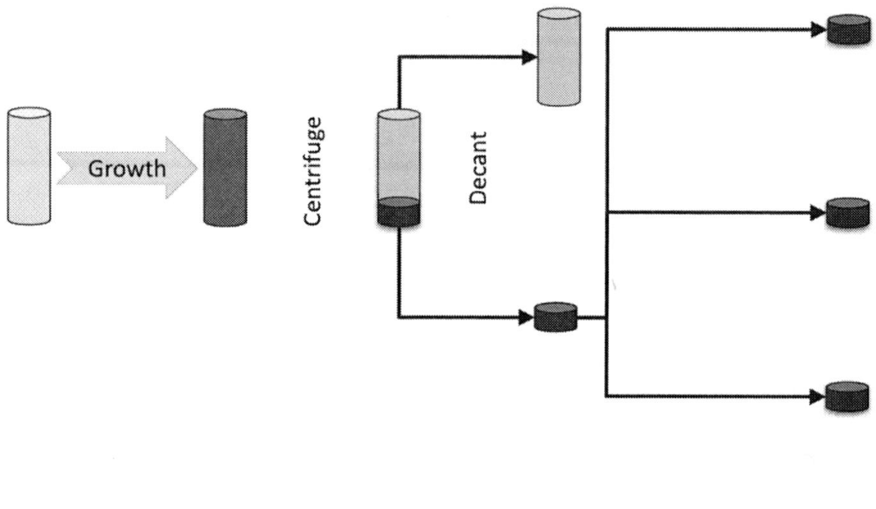

Experimental Design

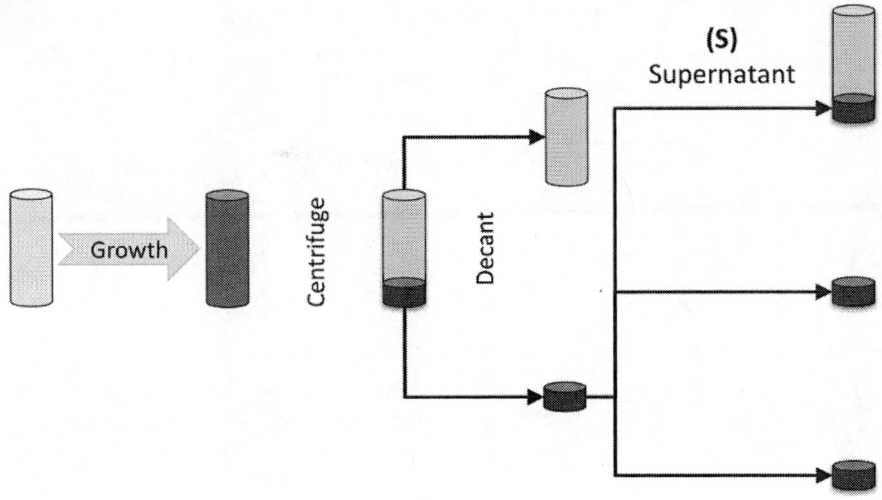

Experimental Design

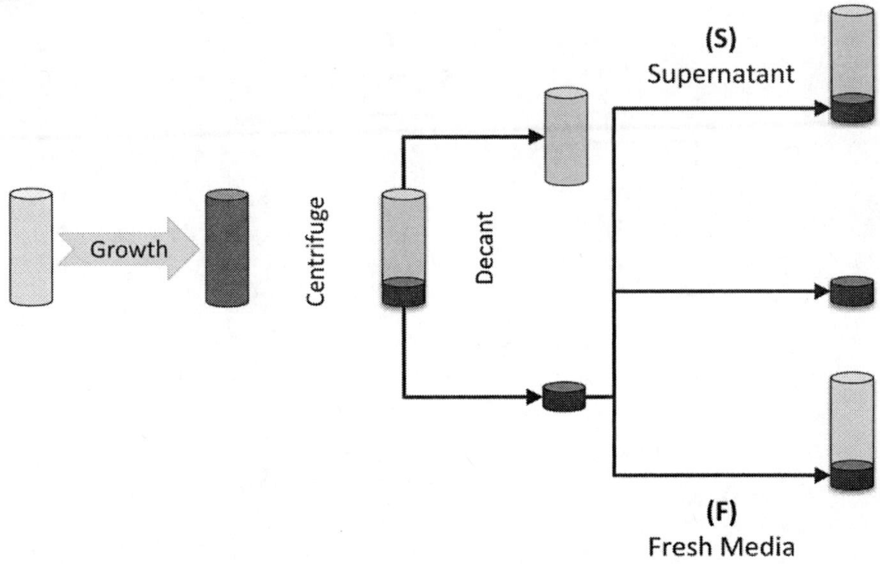

Experimental Design

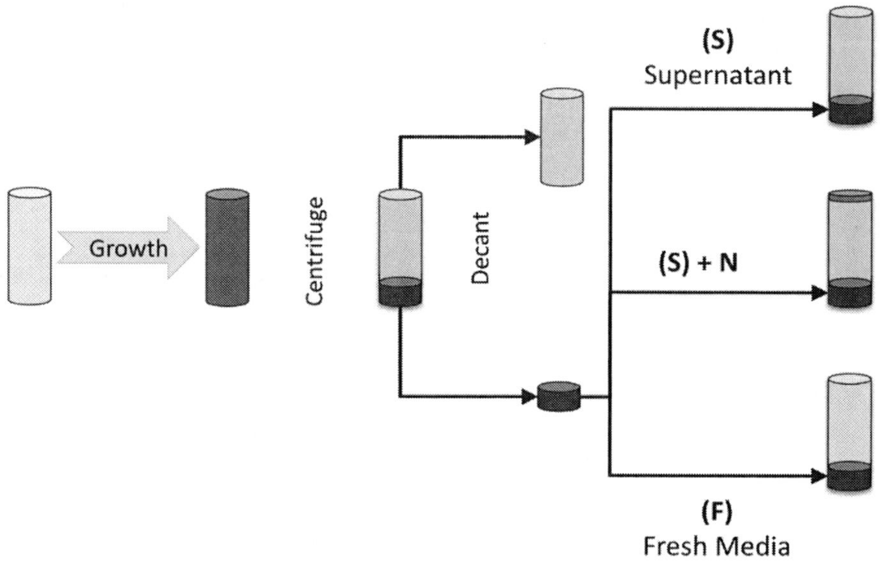

Experimental Design

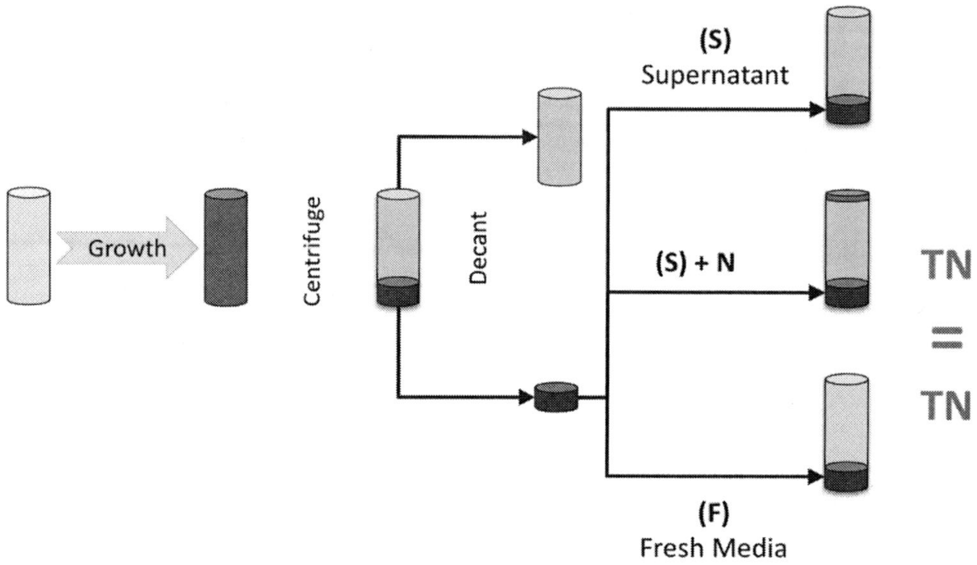

Experimental Design

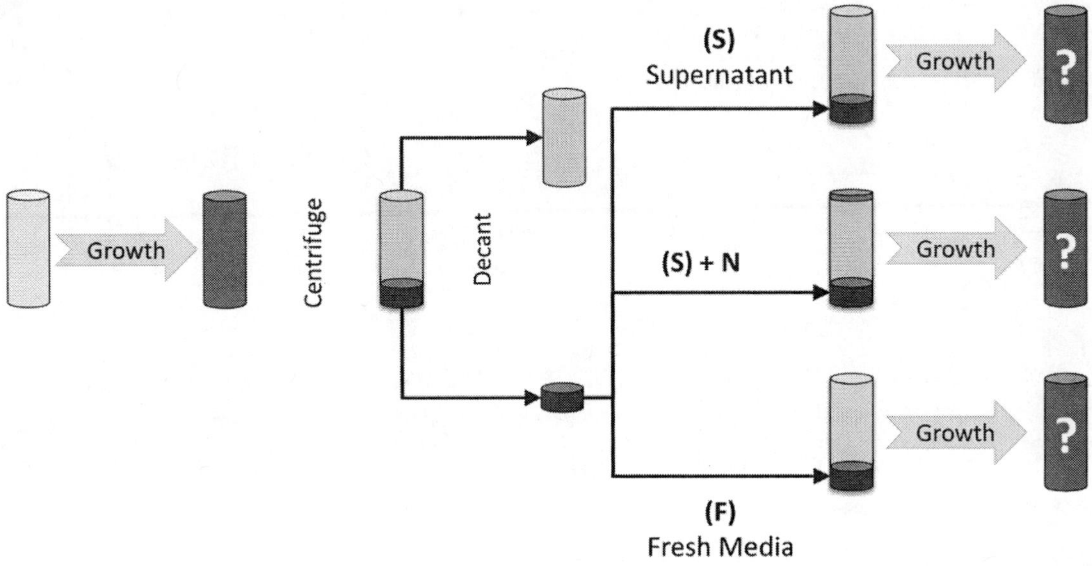

Experimental Design

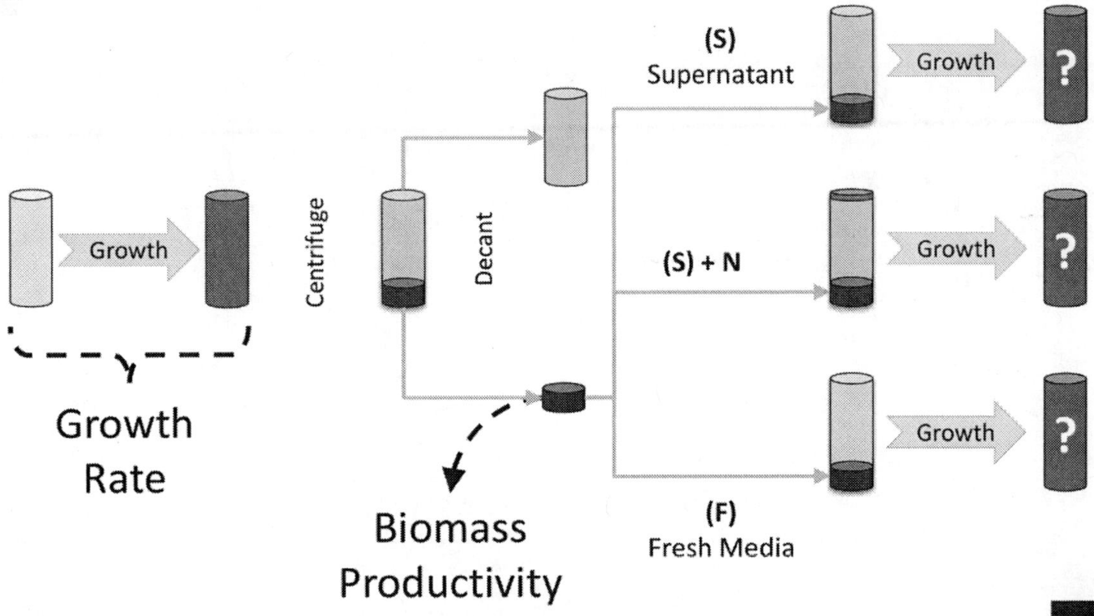

Spirulina platensis Growth

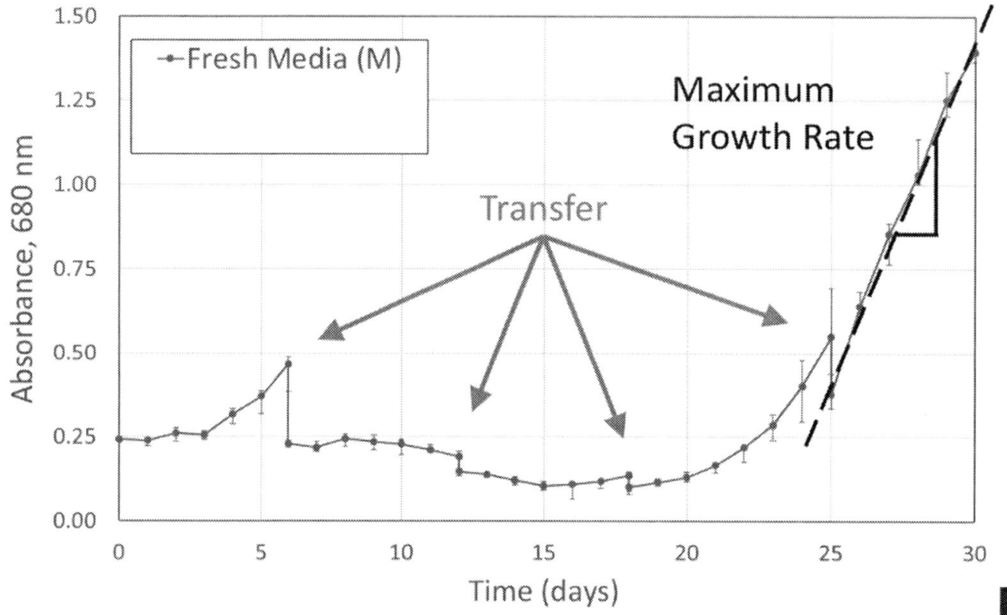

Spirulina platensis Growth

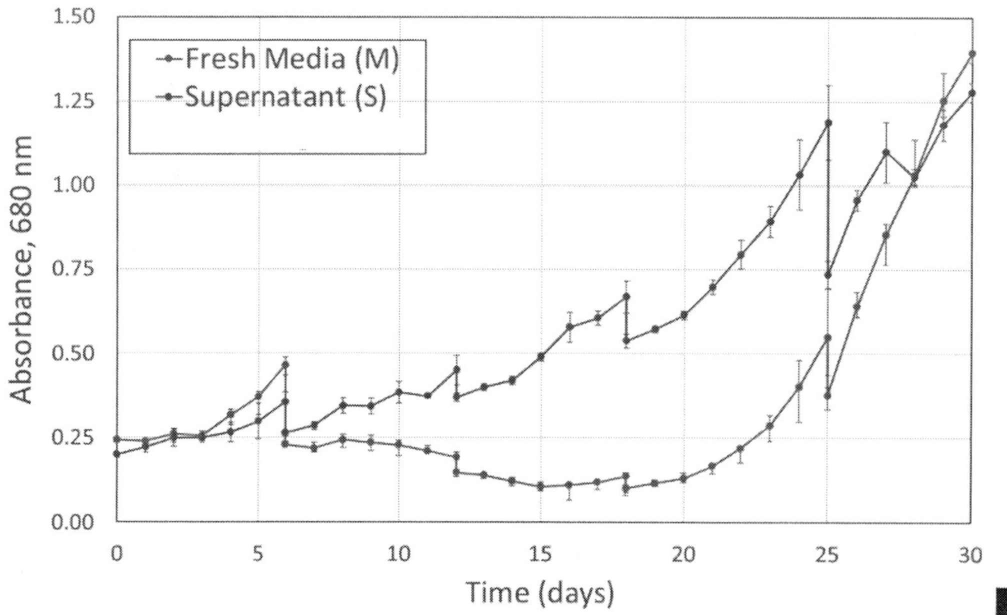

Spirulina platensis Growth

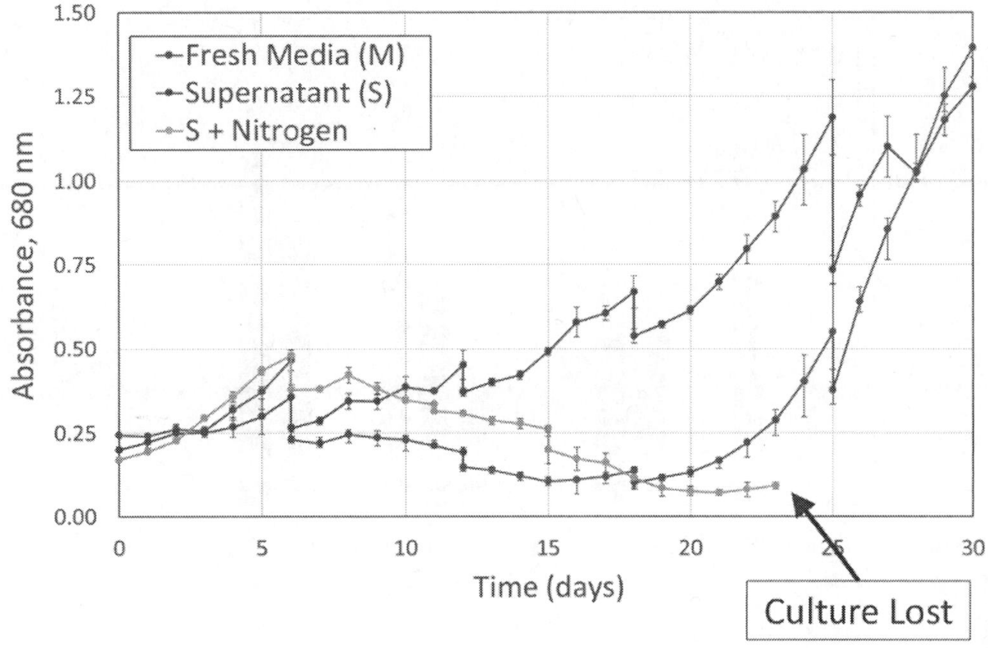

Growth Rates of *S. platensis*

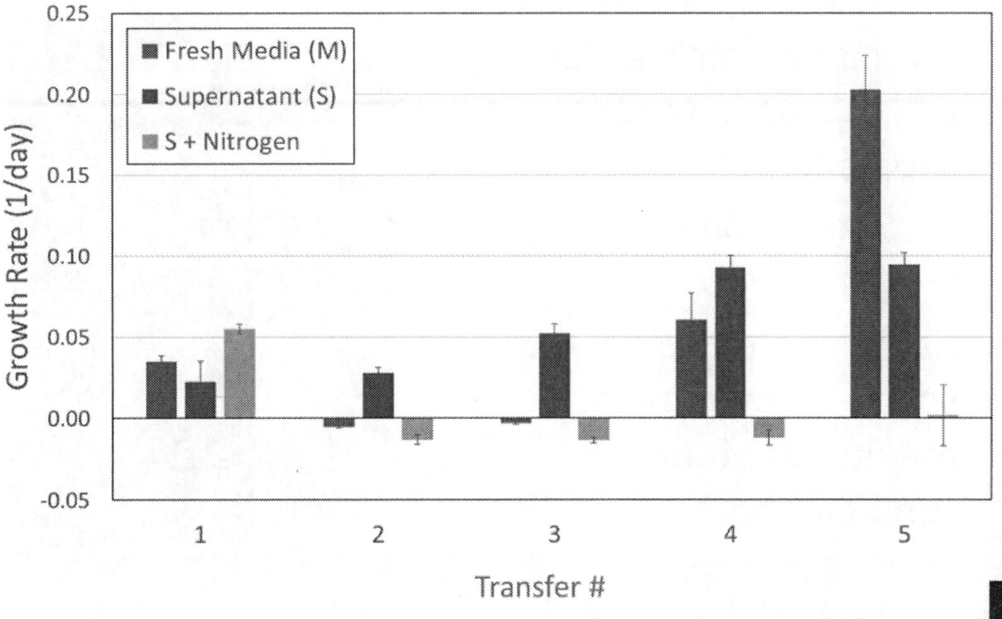

Biomass Productivity

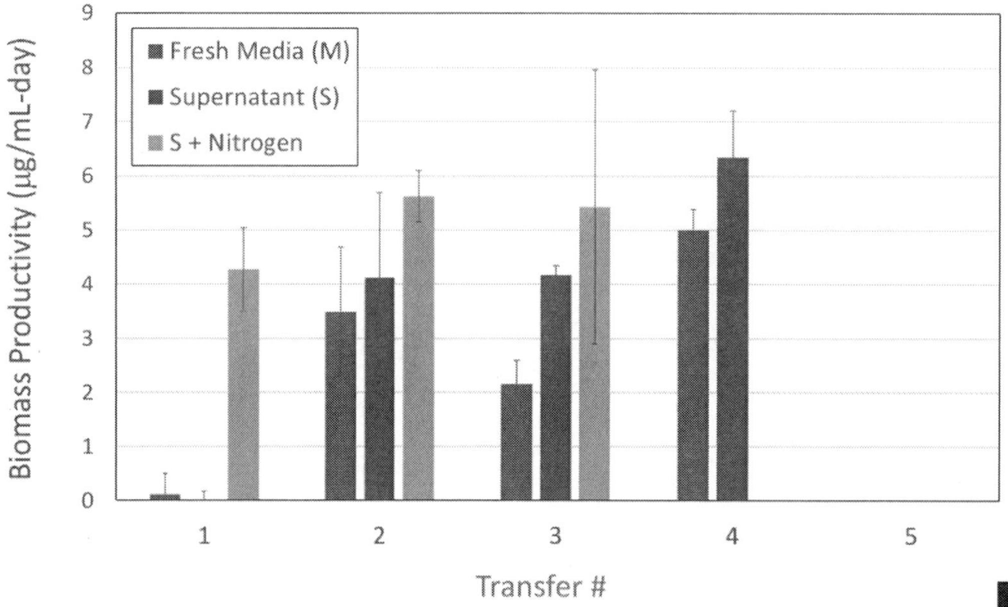

Feed Value

- Crude protein content primary parameter for use as animal feed supplement
  - (S) 47.5%, *dmb*
  - (F) 49%, *dmb*
- Comparable to *Spirulina* species grown fresh medium for food supplements
  - 46 - 64%,*dmb*
    (Lupitini, 2016)

Conclusions

- Reuse of media is valid alternative source of water and nutrients for *S. platensis* cultivation
- Sustained growth rates and biomass productivities observed over multiple reuse cycles
- High crude protein values maintained with reuse media after multiple reuse cycles

Acknowledgments

- Colton Neary
- Richard Mannschreck

- USDA Western Sun Grant Initiative # 2010-38502-21839
- National Science Foundation # 1265075

SunGrant INITIATIVE

NSF

Discussion

Sage R. Hiibel
Chemical and Materials Engineering
University of Nevada, Reno
shiibel@unr.edu

1st - 5th October, 2017
Barcelona, Spain
WCCE10 ECCE11 ECAB4
10th World Congress of Chemical Engineering
Promoters: WCEC EFCE ESBES
Organized by EFCE-Spain Group: ANQUE SEQUI
www.wcce10.org

Towards a more sustainable management of desalination brines

Herrero M.; Dominguez-Ramos A., Ibáñez R., Irabien A.

SOSPROCAN
Department of Chemical and Biomolecular Engineering
Universidad de Cantabria (Spain)

Sosprocan SIPS Integración de Procesos Sostenibles

UC UNIVERSIDAD DE CANTABRIA

1

- Desalination Sustainability Issues
- Case of Study: EDBM
- Experimental Set Up
- Experimental Results
- Conclusions and Future Work

2

Desalination Sustainability Issues

DESALINATION

- **Essential for freshwater supply** in countries under water shortages
- Total worldwide installed **desalination capacity** in 2016
  100 million $m^3 \cdot day^{-1}$ [1]
- **Brine Generation** → **Dumping Areas: Flora and Fauna Affected**
  Posidonia oceanica [2]

- **High Energy Consumption** → **CO_2 Emissions**
  25 kg of oil to produce 1 m^3 of freshwater by desalination [3]

- **More Sustainable Brine Management Required**
- **Challenge:**
  - **Brine Valorization**
  - **Renewable Sources Energies**

[1] Fernandez-Gonzalez C. PhD Thesis 2017.
[2] Liu T et al. Desalination 2013; 326:10-18.
[3] Kalogirou SA. Prog Energ Combust 2005; 31:242–81.

3

Desalination Sustainability Issues

BRINE MANAGEMENT

ADVANCED MANAGEMENT

- Wind Aided Intensified Evaporation (WAIV)
- Membrane distillation
- Forward Osmosis
- Zero Liquid Discharge Systems

Technological Maturity

- Laboratory Level
- Pilot Scale Demonstration
- Industrial Application

BRINES FROM DESALINATION PLANTS

Precipitation	$Mg(OH)_2$	$CaSO_4$ $CaSO_3$
Electrodialysis	HCl	NaOH
Crystallization	NaCl	Na_2SO_4

Pérez-González A et al. Water Res 2012; 46:267-283.

4

Desalination Sustainability Issues

Energy Uses

Reverse Osmosis

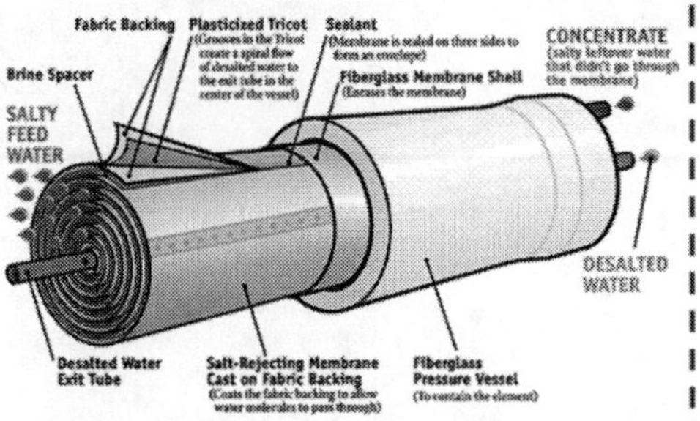

- Pressure Gradient
- Sea water desalination
- 65% desalination capacity*

Electrodialysis

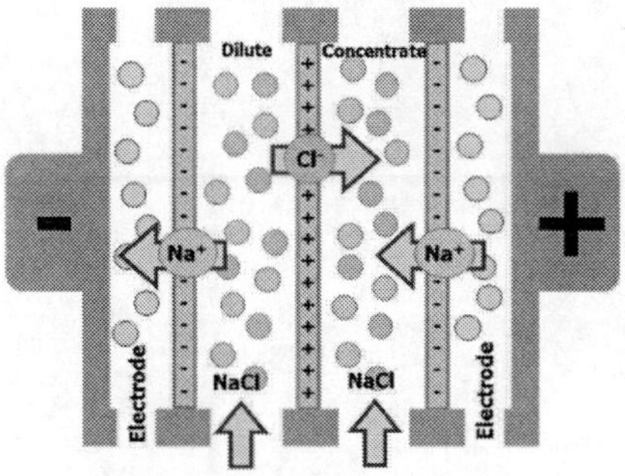

- Electrical Gradient
- Brackish water desalination and brine management
- 3% desalination capacity*

Membrane Processes → ↑ **ENERGY** → **Renewable Sources Energies**

*IDA Yearbook 2014-2015.

Desalination Sustainability Issues

Renewable Energy Desalination

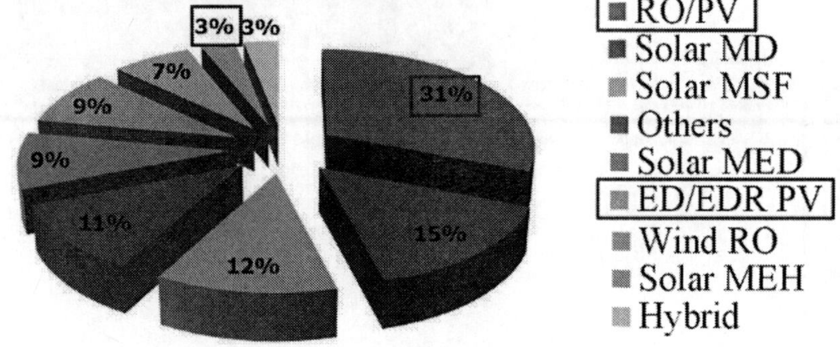

- More than **200** renewable desalination plants installed.
- **1%** of the total installed desalination capacity.

- **34%** of renewable desalination **plants** are powered by PV-Energy.
- **43%** of renewable desalination **capacity** are powered by PV-Energy.

Fernandez-Gonzalez C et al. Renew and Sust Energ Rev 2015; 47: 604-615.

Case of Study: EDBM

State of the art

POWER SUPPLY	GRID 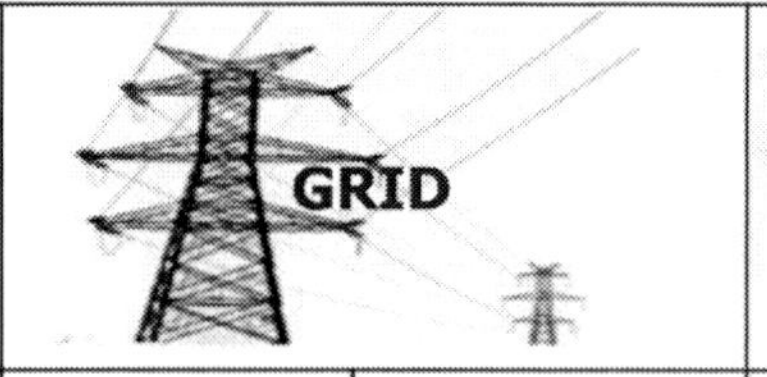		PHOTOVOLTAIC 	
OPERATION MODE	BATCH	CONTINUOUS	BATCH	CONTINUOUS
TECHNOLOGY: ED 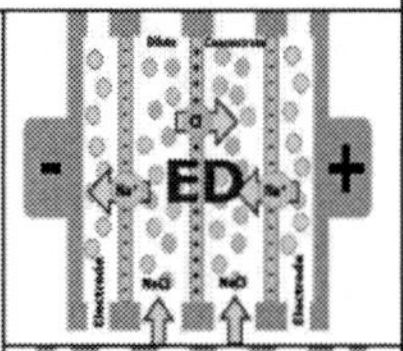	Amor et al., 2001 Tsiakis and Papageorgiu, 2005 Reig et al., 2014	Tanaka et al., 2003 Tsiakis and Papageorgiu, 2005 Fidaleo and Moresi, 2011 Reig et al., 2014	Ortiz et al., 2007 Ortiz et al., 2008 Fernandez-Gonzalez et al., 2015	
TECHNOLOGY: EDBM 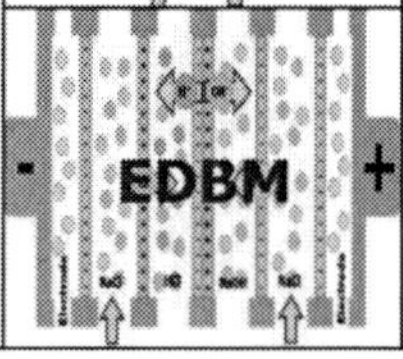	Mier et al., 2008 Fernandez-Gonzalez et al., 2016 Reig et al. 2016	**THIS WORK** Contribution to the desalination (SWRO) sustainability by the integration of PV solar energy to a EDBM brine management process for acid and base production		

Amor Z et al. Desalination 2001; 133(3): 215-223.
Fernandez-Gonzalez C et al. Renew and Sust Energ Rev 2015; 47: 604-615.
Fernandez-Gonzalez C et al. Sep Purif Technol 2016; 45(4): 275-287.
Fidaleo M and Moresi M. J Membrane Sci 2011; 367(1-2): 220-232.
Mier M et al. Sep Purif Technol 2008; 59(2): 197-205.
Reig M et al. Desalination 2014; 342: 107-117.
Reig M et al. Desalination 2016; 382: 13-20.
Ortiz J et al. Desalinaton 2007; 208: 89-100.
Ortiz J et al. Sol Energ Mat Sol C 2008; 92(12): 1677-1688.
Tanaka Y et al. J Membrane Sci 2003; 222(1-2): 71-86.
Tsiakis P and Papageorgiou LG. Desallination 2005; 173:173-186.
Yang Y et al. J Membrane Sci 2014; 449: 119-126.

7

Case of Study: EDBM

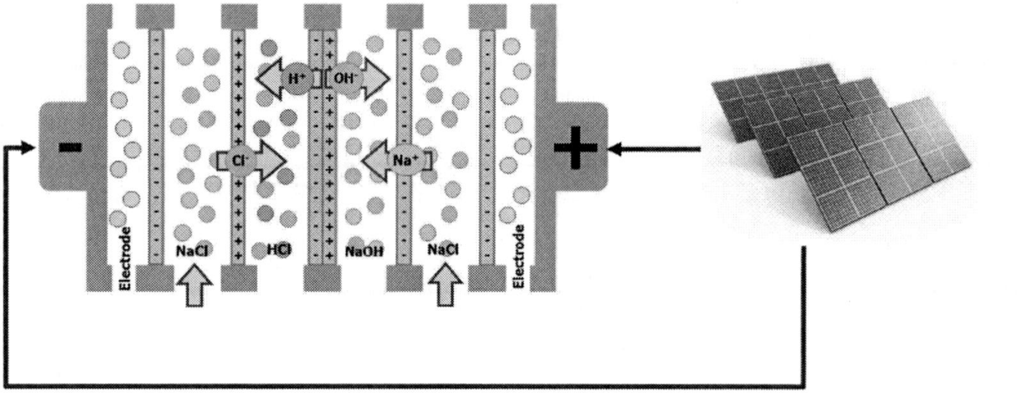

NaCl and H_2O (BRINES) + ↑ **ENERGY** → **HCl and NaOH**

↓

↑ **SEC** ↑ **CO_2 Emissions**

8

Case of Study: EDBM

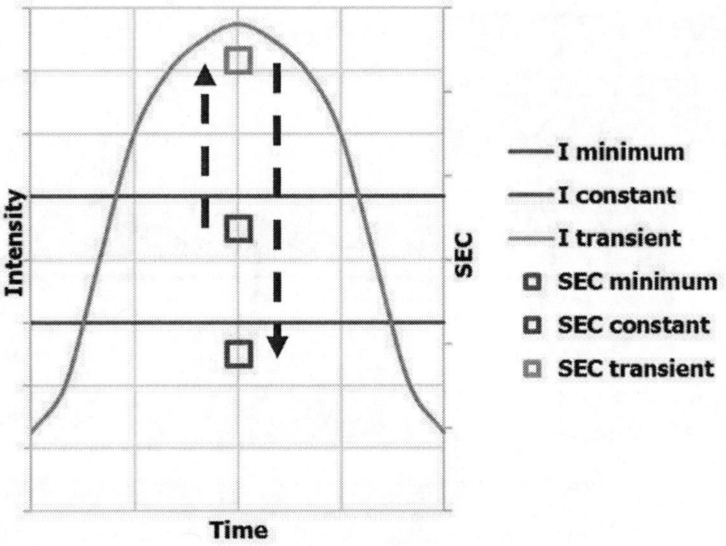

STEP 1 ↓ CO_2 Emissions ➡ PV Energy (Transient) ➡ ↑ SEC

STEP 2 ↓ SEC ➡ Use of Control Loops

Experimental Set Up

Solar Simulator

Electrodialysis cell
A = 100 cm²

Commercial membranes:
- RALEX (CM-PP AM-PP)
- FUMASEP-FBM

Power supply
0-36 V, 0-22 A

Refrigeration bath
$T_{working}$ = 20 °C

4 Feed tanks
V = 2 L

2 Peristaltic pumps

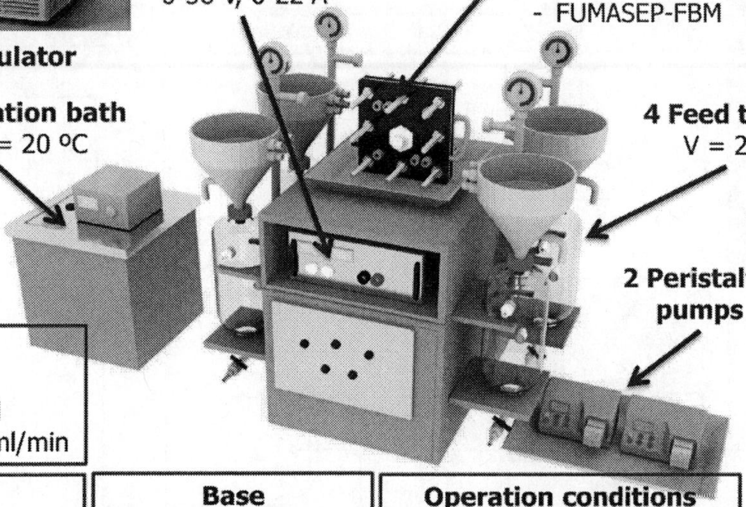

Electrode
[NaOH] or [Na_2SO_4] = 0.5 M

Brine
$[NaCl]_{in}$ = 1 M
$[NaCl]_0$ = 0.5 M
$Q_{in} = Q_{out}$ = 2 ml/min

Acid
$[HCl]_0$ = 1 M
$Q_{in} = Q_{out}$ = 1 ml/min
$[HCl]_{in}$ = 0.1 M

Base
$[NaOH]_0$ = 1.5 M
$Q_{in} = Q_{out}$ = 1 ml/min
$[NaOH]_{in}$ = 0.1 M

Operation conditions
Q = 60 l/h
J = 220 A/m² and Solar Irradiation in July-Almería

Experimental Set Up

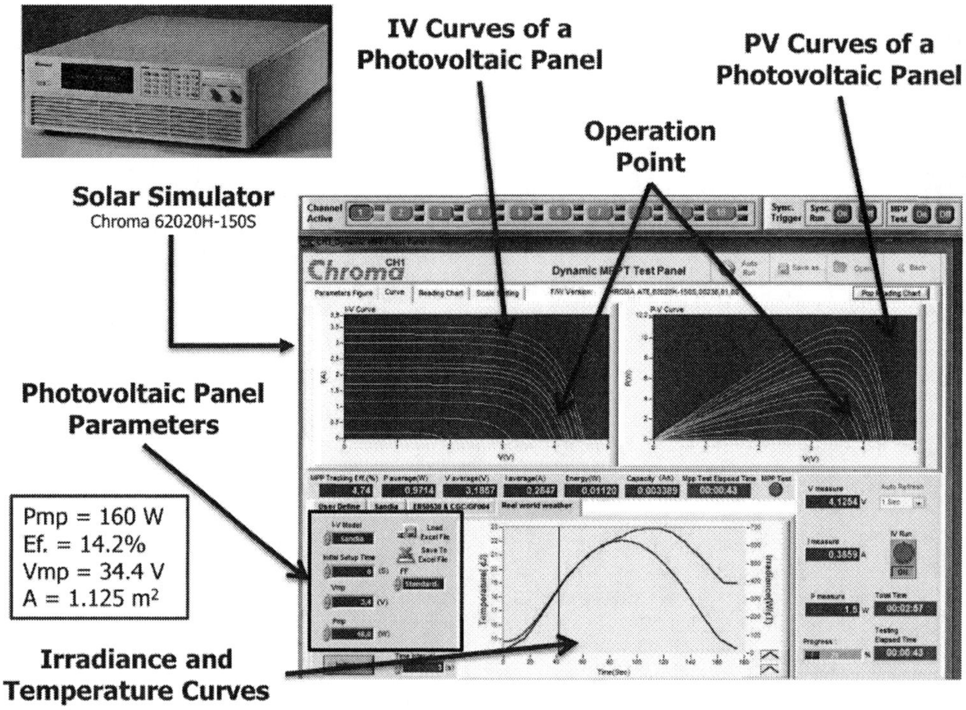

11

Experimental Set Up

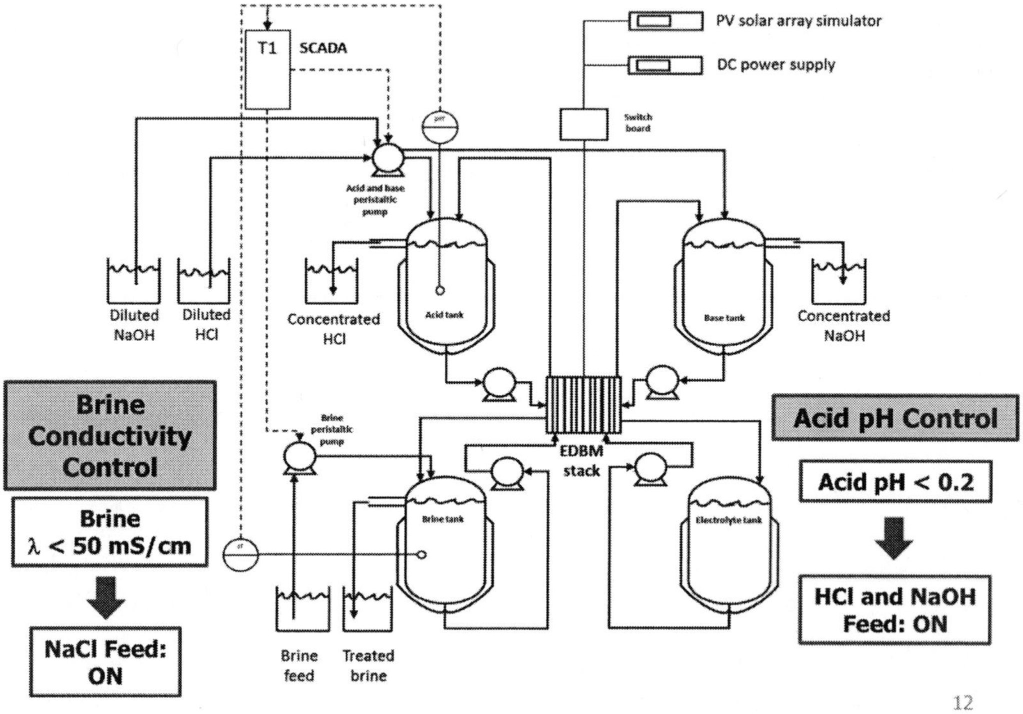

12

Experimental Set Up

Experimental Conditions

Constant Energy

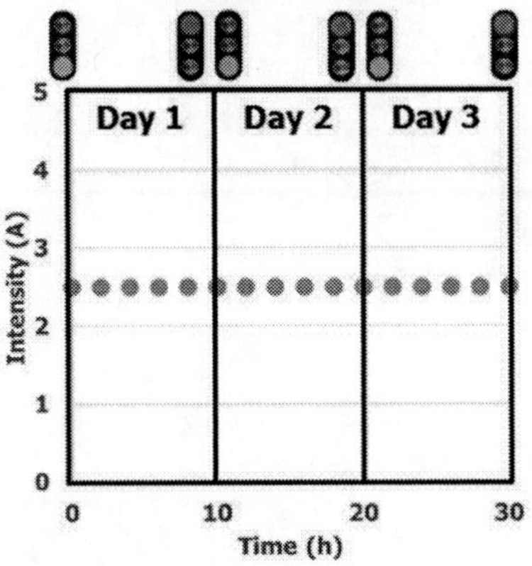

J = 220 A/m² from Grid Mix

Ibañez R et al. Desalination 2013; 309: 165-170.

Transient Energy

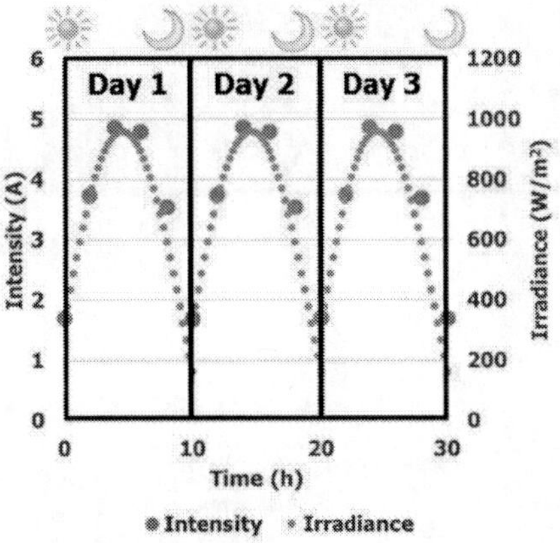

Simulated Solar Irradiation:

July-Almeria

PVGIS - http://re.jrc.ec.europa.eu/pvgis/apps4/pvest.php

13

Experimental Set Up

Experimental Conditions

Exp. code	Current intensity mode	Operation mode	Flowrate (salt tank) (mL·min⁻¹)	Flowrate (acid/base tank) (mL·min⁻¹)	Electrode solution
E-C	Constant	Continuous	2	1	NaOH
E-T1	Transient				
E-T2					Na_2SO_4
E-T3		Semi-continuous	8	4	

14

Experimental Results

Outflow Steady-State Concentrations

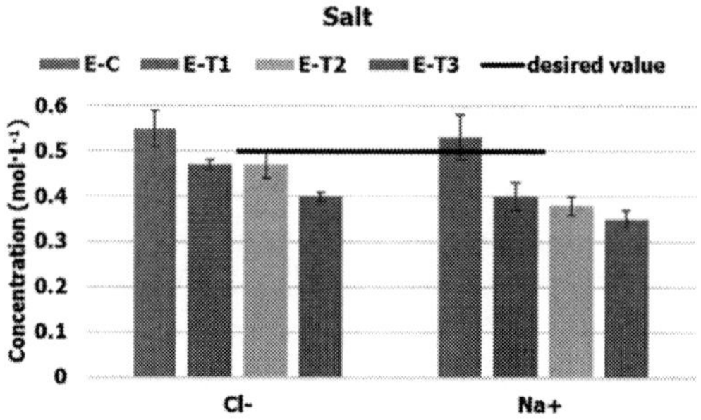

1 M → 0.5 M

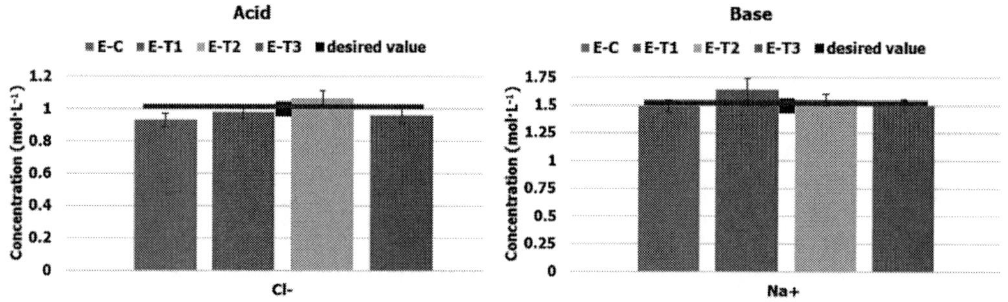

0.1 M → 1 M

0.1 M → 1.5 M

15

Experimental Results

Specific Energy Consumption (SEC)

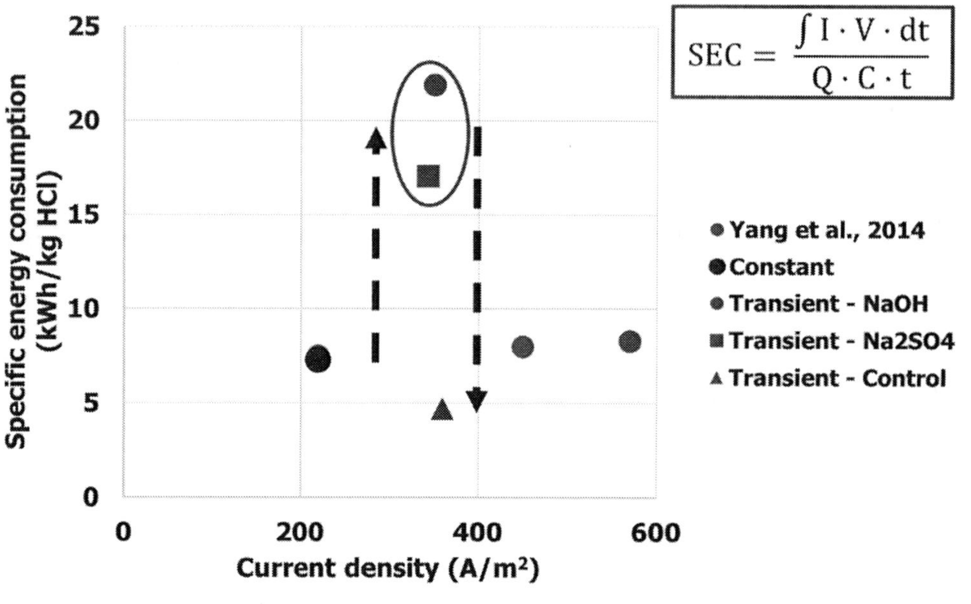

$$SEC = \frac{\int I \cdot V \cdot dt}{Q \cdot C \cdot t}$$

16

Experimental Results

Efficiency of the simulated PV Solar Array

$$\eta\,(\%) = \frac{\int I \cdot V \cdot dt}{A \cdot \int G \cdot dt} \cdot 100$$

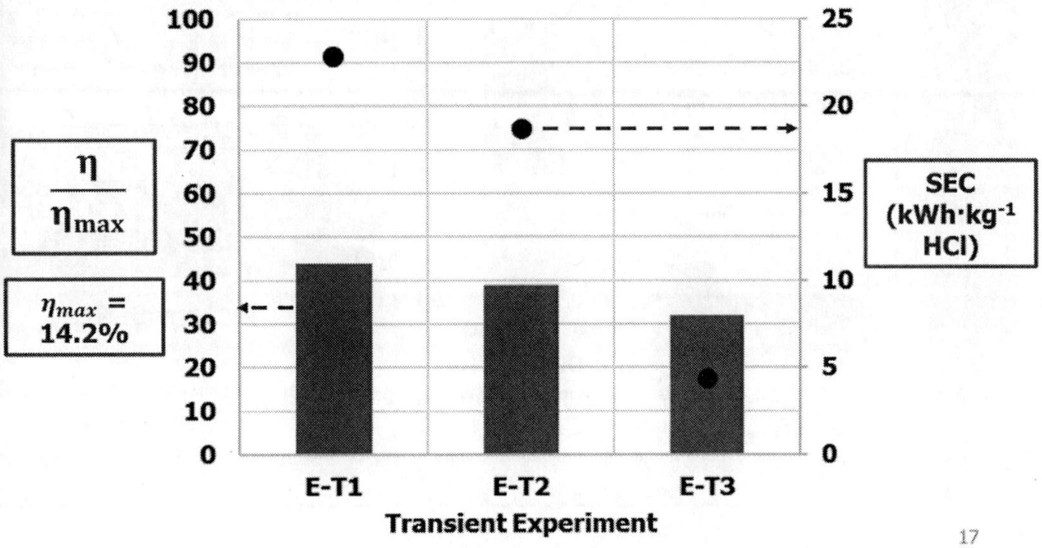

17

Conclusions and Future Work

Conclusions

- ✓ Performance of the process tested using:
  - ✓ Constant current intensity (reference)
  - ✓ Transient current intensity (PV solar array simulator)
- ✓ 1 $mol \cdot L^{-1}$ HCl has been obtained in all experiments
- ✓ Reduction of the SEC from 7.3 $kWh \cdot kg^{-1}$ HCl to 4.4 $kWh \cdot kg^{-1}$ HCl
  - ✓ Transient current intensity
  - ✓ Control loops

18

Conclusions and Future Work

On going

Process Improvements

MPPT to improve the **PV efficiency**

$$\uparrow \frac{\eta}{\eta_{max}}$$

Modeling of the process for optimization

Sustainability evaluation

SCE

המכללה האקדמית להנדסה ע"ש סמי שמעון

Research Stay
Prof. Adi Wolfson
Sami Shamoon College of Engineering
(Be'er Sheva, Israel)

19

1st-5th October, 2017 Barcelona, Spain

10th World Congress of Chemical Engineering

www.wcce10.org

UC UNIVERSIDAD DE CANTABRIA

Acknowledgments:

GOBIERNO DE ESPAÑA — MINISTERIO DE ECONOMÍA Y COMPETITIVIDAD

BES-2015-073501 (CTM2014-57833-R)

Marta Herrero González

Department of Chemical and Biomolecular Engineering
ETSIIyT, Avda. los Castros s/n
Santander, Cantabria. SPAIN
Tel. +34 942 202028 herrerogma@unican.es

osprocan SIPS Integración de Procesos Sostenibles

«We know what we have to do to achieve the sustainability, now it's time to do it »

20

Experimental Results

Concentrations, pH and Conductivity

Exp. code	Steady-state concentration					
	Cl^- ($mol \cdot L^{-1}$)		Na^+ ($mol \cdot L^{-1}$)			SO_4^{2} ($mol \cdot L^{-1}$)
	Salt	Acid	Salt	Base	Electrode	Electrode
E-C	0.55±0.04	0.93±0.04	0.53±0.05	1.49±0.05	0.39±0.09	-
E-T1	0.47±0.01	0.98±0.04	0.40±0.03	1.64±0.10	0.24±0.20	-
E-T2	0.47±0.03	1.06±0.05	0.38±0.02	1.55±0.05	0.32±0.03	0.23±0.01
E-T3	0.40±0.01	0.96±0.05	0.35±0.02	1.50±0.05	0.40±0.01	0.47±0.02
	< 0.5 M	1 M	< 0.5 M	1.5 M	0.5 M	0.5 M

Exp. code	pH				Conductivity ($mS \cdot cm^{-1}$)			
	Salt	Acid	Base	Electrode	Salt	Acid	Base	Electrode
E-C	1.02±0.11	-0.08±0.04	14.23±0.02	13.66±0.12	62±5	320±15	242±13	73±17
E-T1	1.18±0.08	-0.06±0.03	14.24±0.03	13.27±0.46	56±4	349±16	263±12	47±34
E-T2	1.12±0.12	-0.08±0.05	14.24±0.03	0.99±0.24	62±7	367±19	270±14	41±4
E-T3	1.32±0.15	0.03±0.04	14.19±0.01	0.88±0.37	43±1	319±15	243±17	64±8

21

Sustainability Evaluation

Research Stay at SCE

- ✓ SWRO vs SWRO+EDBM
- ✓ 3 different energy scenarios:
  - ✓ Israeli grid mix
  - ✓ Spanish grid mix
  - ✓ PV solar energy
- ✓ Indicators:
  - ✓ Carbon footprint
  - ✓ Water footprint
  - ✓ Land use
- ✓ Normalization and visualization

Santander, Spain

Be'er Sheva, Israel

22

Demand trumps policy:
Bioenergy project experience in South Africa

Professor Harro von Blottnitz (Pr.Eng.)
University of Cape Town
Department of Chemical Engineering
Environmental & Process Systems Engineering Group

International Congress
on Sustainability Science & Engineering '17
Keynote, Session 5: Bringing sustainability to the mainstream

University of Cape Town · iYunivesithi yaseKapa · Universiteit van Kaapstad

UNIVERSITY OF CAPE TOWN

"The global transition has begun—a planetary society will take shape over the coming decades. But its outcome is in question."

Raskin et al. (2002): Great Transition – The Promise and Lure of the Time Ahead; Stockholm Environment Institute

PARIS2015
COP21·CMP11

Are we asking the right questions?

- We, (academic) chemical engineers:
  - "What is the optimal configuration of a sustainable process/product/system?"
    - Analysis - paralysis
- Rather:
  - "How can we help to get low-carbon plants built now?"
    - Action and transition-focused
    - Needs social/economic savvy, not mathematics
    - Flexible platforms, optimised later

UNIVERSITY OF CAPE TOWN

This keynote: Story and analysis

- Contentious role of biofuels in sustainable development
- Bioenergy policy vs. actual projects in South Africa
  - Crop-based liquid fuels
  - Waste-based bioenergy, esp. biogas
- Analysis: Role of product demand
- Conclusions

Biofuels: Potential benefits and risks

- Potential benefits: Biofuels offer attractive features for complementing fossil fuels:
  - Domestic fuel feedstock
  - Carbon neutrality (photosynthesis balances combustion): CAREFUL!!
  - Employment creation potential
- Risks are diverse and often location-specific
  - Land use → biodiversity losses
  - Eutrophication
  - Water use
  - Impacts on food production, availability, prices

UNIVERSITY OF CAPE TOWN

Biofuels: there are more than 2!

- Fuel ethanol & FAME biodiesel
- Also 1st generation:
  - Biogas, charcoal, ethanol gel
- 2nd generation:
  - Cellulosic ethanol, algal biodiesel
  - Methanol, butanol, DME
  - Syngas, pyrolysis oils
  - BtL fuels, hydrogen
- And when we think ‘vehicle fuels’ we need to think super-structure and how its landscape is changing!

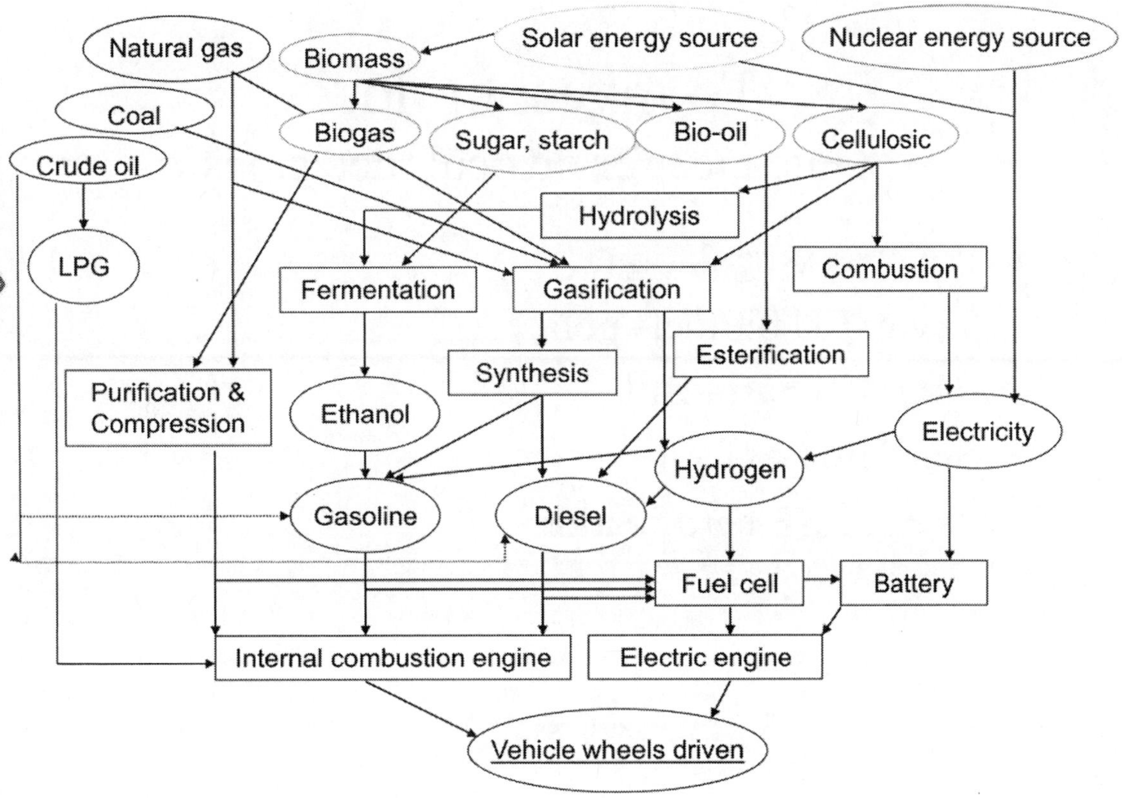

UNIVERSITY OF CAPE TOWN

A significant literature on biofuels

- One fairly consistent finding is that liquid biofuel projects, particularly when made from purpose-grown energy crops or even diverted food crops, have significantly higher sustainability risks than waste-based bioenergy installations.

UNIVERSITY OF CAPE TOWN

To invest or not?

Guidance for difficult decisions

- Policy & Regulation
  - e.g. EU biofuels policy
  - Or US Farm Bill
- Standards
  - e.g. RSB certification

UNIVERSITY OF CAPE TOWN

This analysis

BIOENERGY PROJECT EXPERIENCES IN SOUTH AFRICA

UNIVERSITY OF CAPE TOWN

Timelines

- Mid 2000s: Feasibility study for maize-based bioethanol project
- Late 2007: Industrial Biofuel Strategy adopted by Cabinet
- 2015: Blending targets for bioethanol and biodiesel
- 2017: No biofuel in fuel pool
- 2011: SA hosts COP17
- 2011: start of the REIPPPP*, mainly wind, solar PV, some CSP, minor biogas
- 2013: SABIA formed
- 2014-15: load-shedding
- 2017: > 20 biogas; > 10 LFG installations operational

* renewable energy independent power producer programme

UNIVERSITY OF CAPE TOWN

Key differences

- Liquid biofuels:
  - Strong desire by national government to use as means of rural job creation; strategy approved by cabinet; incentives devised; blending targets set; feedstock would have to be grown
  - Unresolved modernisation of liquid fuels sector
- Waste-based biogas:
  - Not viewed as significant RE contributor; multiple regulatory barriers; feedstock available and with environmental liabilities
  - Waste management law introduced 2008

A business view?

- Why make a fuel that the industry won't buy until they've come to an agreement with government?
- Unmet (energy) demands?

Energy demand?

http://www.bio2watt.com/gallery.html

Resulting projects

http://www.ibert.co.za/

http://cleanenergyafrica.co.za/portfolio-item/new-horizons-energy/

UNIVERSITY OF CAPE TOWN

A closer look

- All projects need at least 2 income streams
  - And thus, meet ≥ 2 demands
- Case of New Horizons Energy:
  1. Demand for industrial fuel gas
     - Imported LPG in Cape Town twice as expensive as pipeline gas in Johannesburg
  2. Demand for waste management services
     - Landfill tariffs in Cape Town 75% higher than in Johannesburg
     - Created by the 2008 Waste Act and 2011 National Waste Management Strategy
- Another key: "good governance" in the public sector

UNIVERSITY OF CAPE TOWN

Conclusions

- Big need for radical change
  - to go mainstream
- Government policy alone won't make change happen
- Business identifying unmet demand and innovating to meet it is a key driver
- Policy that is enabling (but not prescriptive)
  AND good governance are important, too!

- Build flexible low-carbon platforms now,
  - optimise later?

UNIVERSITY OF CAPE TOWN

And then there's (hidden) demand at small scale …

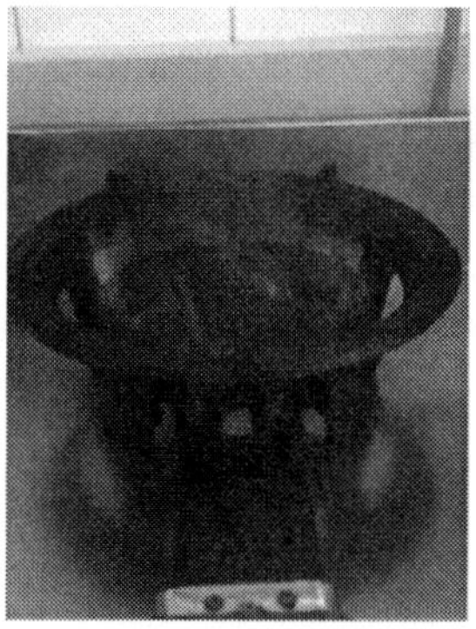

UNIVERSITY OF CAPE TOWN

Acknowledgements

- To the following colleagues who have helped me refine my thinking on this topic:
  - Lauren Basson, Usisipho Gogela (GreenCape); Rethabile Melamu, Maloba Tsehla (The Innovation Hub); Sean Thomas (Bio2Watt); Egmont Otterman (New Horizons Energy)
  - Multiple postgraduate students who have worked on biofuels and biogas projects over the years

UNIVERSITY OF CAPE TOWN

Readings & Bibliography

- Farrell, A.E., R.J. Plevin, B.T. Turner, A.D. Jones, M. O'Hare and D. M. Kammen: *Ethanol can contribute to energy and environmental goals*, Science 311, pp. 506-508, 2006.
- Fargione, J, J Hill, D Tilman, S Polasky, P Hawthorne: Land Clearing and the Biofuel Carbon Debt; SCIENCE, 319,1235-1238, 2008
- Kaltschmit, M, Reinhardt, GA, & Telzer, T; *Life cycle analysis of biofuels under different environmental aspects*, Biomass & Bioenergy, 12, pp. 121-134, 1997.
- Letete, T. and H. von Blottnitz; Biofuel Policy in South Africa: A Critical Analysis; in Bioenergy for Sustainable Development in Africa; Janssen, R. and D. Rutz (eds); Springer, ISBN 978-94-007-2180-7, 2011.
- Melamu, R. and H. von Blottnitz: 2nd Generation Biofuels a Sure Bet? A Life Cycle Assessment of How Things Could Go Wrong; *Journal of Cleaner Production*, 19 (2-3), pp. 138-144, 2011.
- Melamu, R. and H. von Blottnitz; The Development of a Commercial Biogas Project in Bronkhorstspruit, South Africa: A Review; Report for Bio2Watt, 2013.
- Melamu, R. and H. von Blottnitz: The Impact of a Knowledge Diffusion Event on the Development of the Nascent Biogas Industry in South Africa; 5th International Conference on Sustainable Transitions, Utrecht, The Netherlands, 27-29 August, 2014.
- Nasterlack, T. von Blottnitz, H. and R. Wynberg: Are biofuel concerns globally relevant? Prospects for a proposed pioneer first-generation bioethanol project in South Africa, *Energy and Sustainable Development*, 23, pp. 1-4; 2014.
- von Blottnitz, H. and M.A. Curran; *A review of assessments conducted on bioethanol systems from an energy balance, CO_2, and environmental life-cycle perspective*; Journal of Cleaner Production, 15 (7), pp 607-619, 2007.

UNIVERSITY OF CAPE TOWN

Completed dissertations

- R. Melamu: "Waste-based bioenergy in Africa: Operationalizing technology innovation systems analysis to go beyond assessments of potential", PhD, 2014.
- M. Robinson: "The role of social capital in the design and management of waste-to-energy anaerobic digestion infrastructure"; UIDM MPhil, 2013 .
- G. Munganga: "Bioprocessing Technologies for Municipal Solid Waste in African Cities", MSc, 2012.
- I. Dalwai, "A Comparison of Technical and Environmental Merits of Producing Bio-Ethanol and Bio-Methane from Waste Paper Sludge", MSc, 2012.
- C. Trautmann: "Investigation of the Use of Biogas in a Gas Hob and the Feasibility of Upgrading it on a Household Scale", MSc, 2012.
- L. Malla:"Greenhouse Gas Mitigation Cost of Energy From Biogas: A Techno-Economical Analysis of Co-Digestion Of Three Types of Waste", MSc, 2011.
- B. Amigun: "Processing Cost Analysis of the African Biofuels Industry with Special Reference to Capital Cost Estimation Techniques"; PhD, 2008.
- C. Nissing: "Creative deployment of technology in urban planning for sustainable energy use and supply in a South-North comparison", PhD, 2007.
- J. Cohen: "A Life Cycle Assessment into energy recovery from organic waste: A case study of the Water Treatment Facility of SABMiller Newlands Brewery", MSc, 2006.

Hybridization of integrated model used for sustainability assessment

Hanumante, Neeraj C. [1,2,3,#], Shastri, Yogendra [2], Hoadley, Andrew [3]

1 IITB-Monash Research Academy, Mumbai, India
2 India Institute of Technology, Mumbai, India
3 Monash University, Melbourne, Australia
The author is a recipient of Prime Minister's Fellowship Scheme for Doctoral Research, a public-private partnership between Science & Engineering Research Board (SERB), Department of Science & Technology, Government of India and Confederation of Indian Industry (CII). The author's host institute for research is IITB-Monash Research Academy, and the partner company is Tata Chemicals Limited.

1

International Congress on Sustainability Science and Engineering (ICOSSE '17)

IITB-Monash Research Academy
An Indian-Australian research partnership
MONASH University

Sustainability assessment

Sustainable development: "development that meets the needs of the present without compromising the ability of future generations to meet their own needs."

Sustainability Assessment: Assistance in decision making related to actions that may impact sustainability

Tools:

- Indicators, and indices
- Product related assessments
- Integrated assessment[1]
  - ....
  - System dynamics (SD)
  - Agent based modeling and simulation (ABMS)
  - Coupled component modeling or hybrid modeling

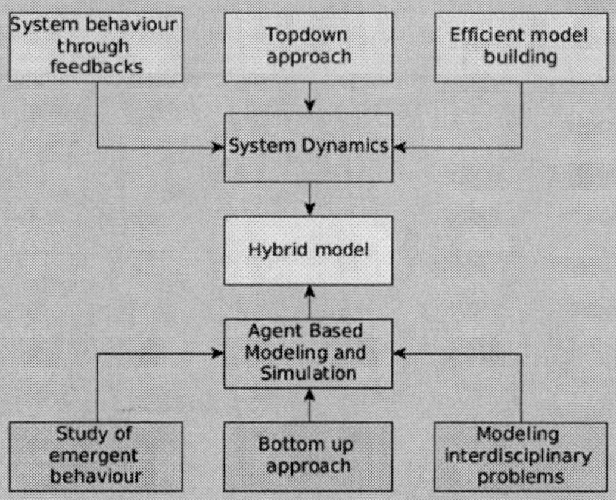

1 - Kelly (Letcher) 2013

2

14-compartment model[2,3,4,5,6,7]

- Closed to mass, and energy
- Food web with multiple trophic level
- Classification of flora, and fauna as human controlled, and independent
- Socio-economic construct introduced through industries viz. P1, H1, IS, and EP
- Intercompartmental mass flows affected by human activities - P1H2, H1C2, P2H1
- Following mortality mass from each organismic compartment flows back to RP, very slow recycling flow from IRP to P2, and P3

2 - Cabezas et al 2003
3 - Fath and Cabezas 2004
4 - Cabezas et al 2005
5 - Whitmore et al 2006
6 - Shastri et al 2008
7 - Kotecha et al 2010

Hybridization

- Compartment selection[8]
  - modular or self contained
  - capability of decision making
  - possibility of interaction
- **Human compartment** selected for modification
- Modifications:
  - Each human agent is treated as separate entity
  - Human population is divided into affluent, and regular groups.
  - Affluence factor ratio of consumption by affluent group to regular group

Affluence factor	Value
P1	1.064
H1	1.871
IS	2.314
EE	9.651

- Attributes of the human agents
  - Mass
  - Lifespan
  - Affluence
- Creation of human agents:
  - Number of agents, and available mass to create them is determined by system dynamics part of the model to be created.
- Expiry of human agents
  - Human agents expire through two modes:
    - Completion of lifespan
    - Deviation of mass from ideal

8 - Macal and North 2009

Hybridization - Attributes - Mass

- Initial condition:
  - Available initial mass of human compartment distributed among two groups based on affluence factors
  - Among agents these groups, the mass is distributed based on coefficient of variation selected for that group

Coefficient of variation	Value
Regular	0.37
Affluent	0.33

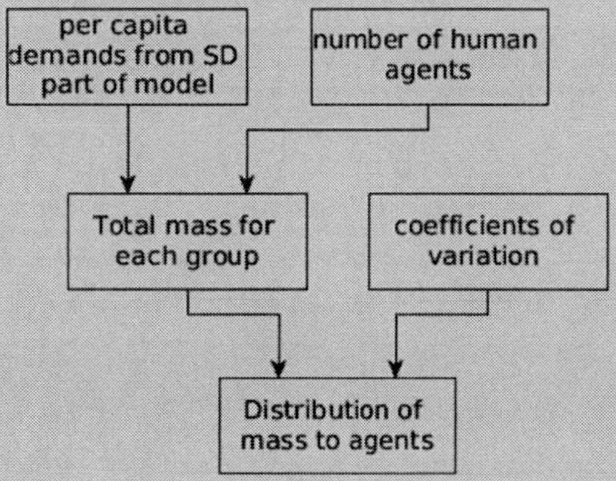

Hybridization - Attributes - lifespan

- Initial condition - Current age at t=0:
  - Birth time allocation is based on population pyramid representing that group
- Life span
  - Life spans are provided based on cumulative frequency of mortality for each age group
  - A frequency is randomly chosen from representative curve, and mapped back to life span in human years.

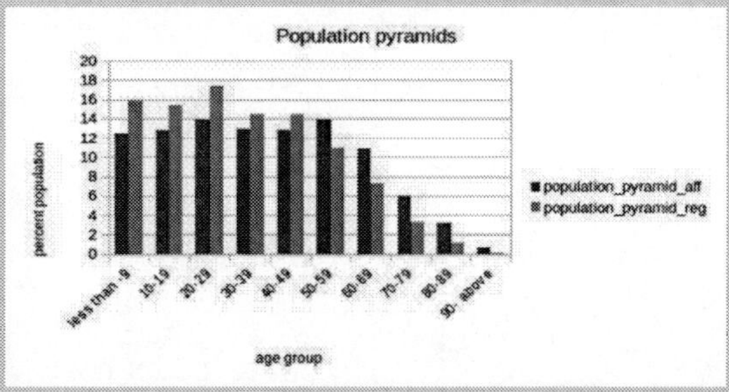

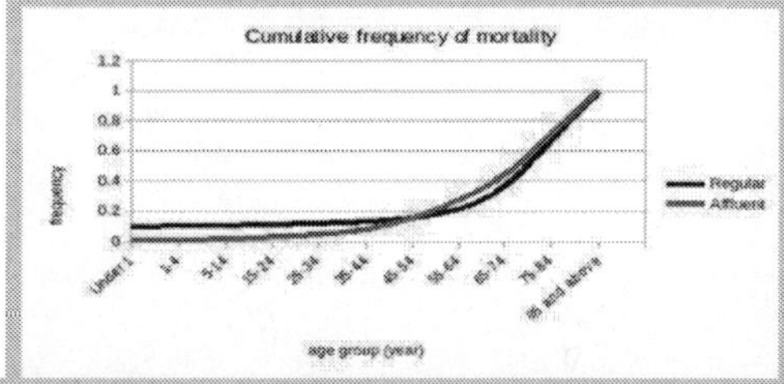

Comparison of hybrid model results with SD model

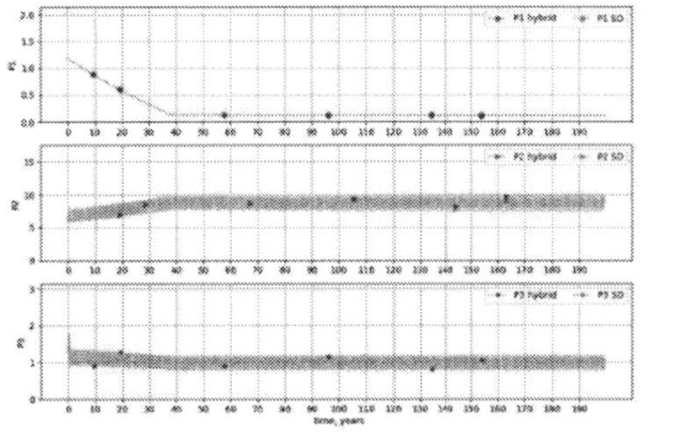

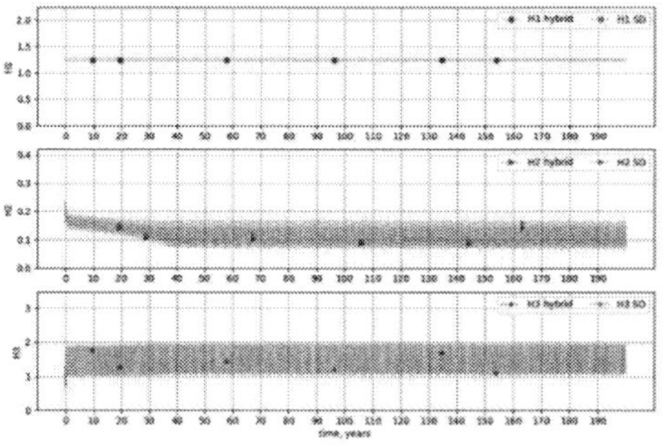

International Congress on Sustainability Science and Engineering
(ICOSSE '17)

IITB-Monash Research Academy
An Indian-Australian research partnership

MONASH University

Comparison of hybrid model results with SD model

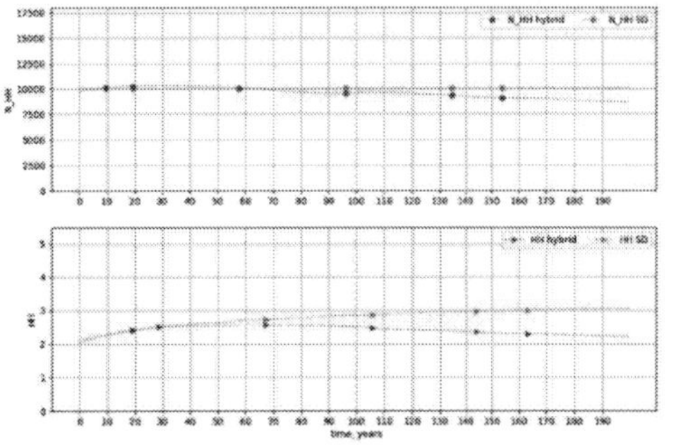

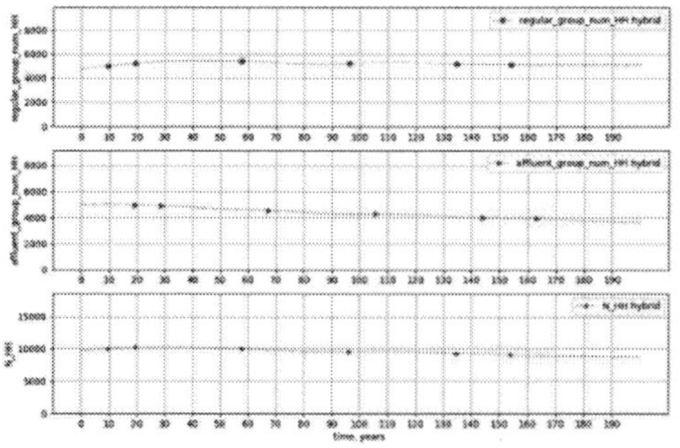

International Congress on Sustainability Science and Engineering (ICOSSE '17)

IITB-Monash Research Academy

An Indian-Australian research partnership

Testing the model for variation of regular group fraction

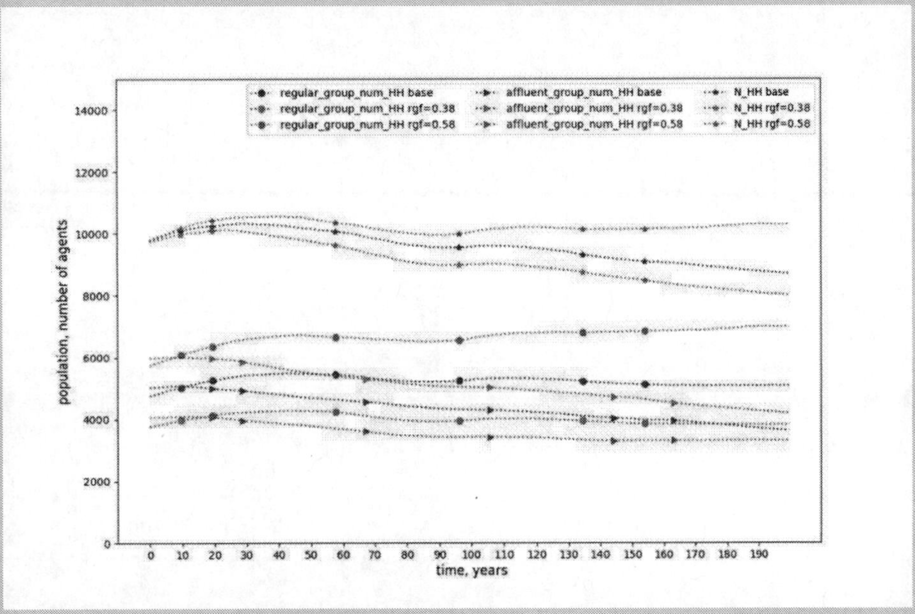

9

Scenarios to be modelled using hybrid model

The animal farming

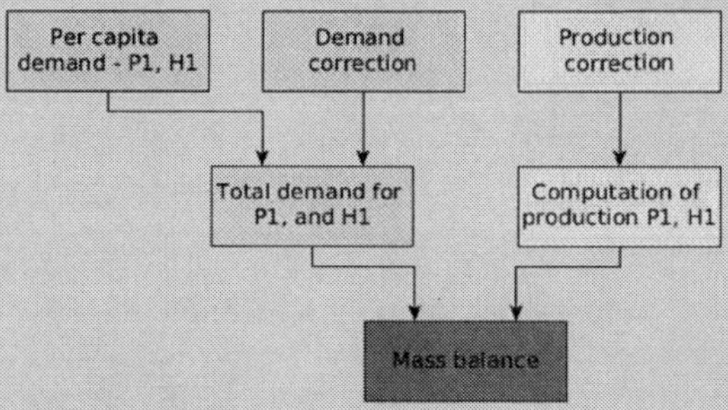

Reduction in animal based products

- Regular group fraction = 0.48
- Two routes:
  - Through demand correction
  - Through production correction
- Two cases
  - Without substitution by plants
  - With substitution by plants

Image: Modern animal husbandry
by : Graeme Churchard, Taken on March 22, 2009
Attribution 2.0 Generic (CC BY 2.0)

International Congress on Sustainability Science and Engineering
(ICOSSE '17)

IITB-Monash Research Academy
An Indian-Australian research partnership
MONASH University

Overpopulation

- Simulated by increasing life expectancy of the agents
- Difference in life expectancy increase rate is expected to result in different rates of changes in population of regular and affluent groups
- Overpopulation is expected to strain the resources and may lead to extinction of some species identified by compartment mass becoming zero

Image: Crowd
by : James Cridland, Taken on June 23, 2007
Attribution 2.0 Generic (CC BY 2.0)

Excessive resource consumption

- Simulated by increasing consumption coefficients
- Difference in life expectancy increase rate is expected to result in different rates of changes in population of regular and affluent groups
- Similar to overpopulation it is expected to strain the resources and may lead to extinction of some species identified by compartment mass becoming zero.
- Effects on economic aspects may also be interesting to study.

Image: dutch supermarket AH (8)
by: bertknot, Taken on June 23, 2007

Summary

- Integrated socio-econo-ecological model has been hybridised with ABMS.
- The current form of the model has been tested against variation of initial regular group fraction and has been found to be stable.
- Scenarios considered to be modeled for expansion of current form of the model and explore its behaviour:
  - Animal farming scenario
  - Excessive consumption of resources
  - Population explosion
- Further, in the current form of the model, lifespan and mass allocated to an agent, are independent. This aspect may be altered in future.

14

Acknowledgement

This research has been aided by Prime Minister's Fellowship Scheme for Doctoral Research, a public-private partnership between Science & Engineering Research Board (SERB), Department of Science & Technology, Government of India and Confederation of Indian Industry (CII).

We gratefully acknowledge the support of our industry partner: Tata Chemicals Ltd.

We thank in particular Alka Talwar, Chief CSR & Sustainability Officer for her support to the project.

15

International Congress on Sustainability Science and Engineering (ICOSSE '17)

IITB-Monash Research Academy
An Indian-Australian research partnership

References

1. **Kelly (Letcher)**, Rebecca A., Anthony J. Jakeman, Olivier Barreteau, Mark E. Borsuk, Sondoss ElSawah, Serena H. Hamilton, Hans Jørgen Henriksen, Sakari Kuikka, Holger R. Maier, Andrea Emilio Rizzoli, Hedwig van Delden, and Alexey A. Voinov, **2013** sep, "Selecting among five common modelling approaches for integrated environmental assessment and management," Environmental Modelling & Software 47, 159–181
2. **Cabezas**, H., Pawlowski, C., Mayer, A., & Hoagland, N. T. (**2003**). Sustainability: ecological, social, economic, technological, and systems perspectives. Clean Technologies and Environmental Policy, 5(3), 167–180. http://doi.org/10.1007/s10098-003-0214-y
3. **Fath**, B. D., & **Cabezas**, H. (**2004**). Exergy and Fisher Information as ecological indices. Ecological Modelling, 174(1–2), 25–35. http://doi.org/10.1016/j.ecolmodel.2003.12.045
4. **Cabezas**, H., Pawlowski, C. W., Mayer, A. L., & Hoagland, N. T. (**2005**). Sustainable systems theory: ecological and other aspects. Journal of Cleaner Production, 13(5), 455–467. http://doi.org/10.1016/j.jclepro.2003.09.011
5. **Whitmore**, H. W. J. ., Cabezas, H., Pawlowski, C., Mayer, A., & Hoagland, T. N. (**2006**). Integration of an Economy under Imperfect Competition with a Twelve-Cell Ecological Model Integration of an Economy under Imperfect Competition with a Twelve-Cell Ecological Model. The U.S. Environmental Protection Agency.
6. **Shastri**, Y., Diwekar, U., Cabezas, H., & Williamson, J. (**2008**). Is sustainability achievable? Exploring the limits of sustainability with model systems. Environmental Science and Technology, 42(17), 6710–6716. http://doi.org/10.1021/es800661x
7. **Kotecha**, P. R., Diwekar, U. D., Templeton, J., & Cabezas, H. (**2010**). A Novel Integrated Ecological Model for the study of Sustainability, (2010 IFAC International Symposium on Dynamics and Control of Process Systems July 5-7, 2010, Leuven, Belgium)
8. **Macal**, C. M., & **North**, M. J. (**2009**). Agent-based modeling and simulation. Proceedings of the 2009 Winter Simulation Conference (WSC), 86–98. http://doi.org/10.1109/WSC.2009.5429318

16

Thank you

Additional information

Base case other results

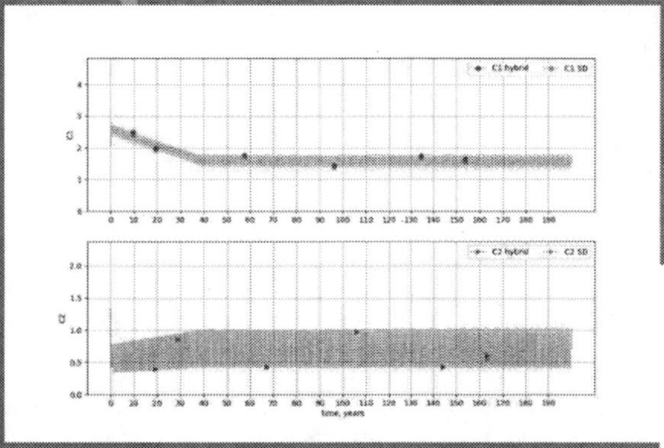

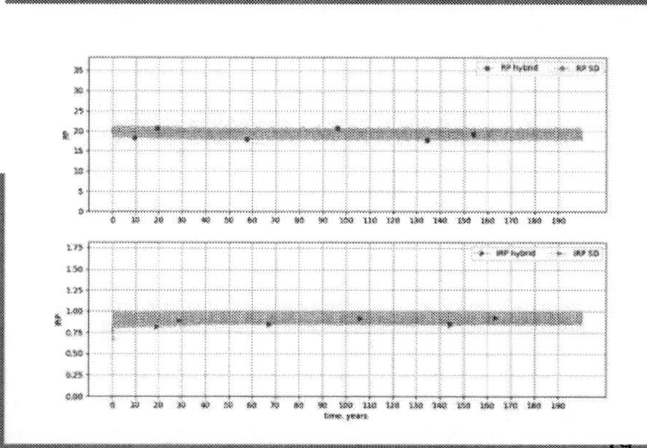

19

International Congress on Sustainability Science and Engineering (ICOSSE '17)
IITB-Monash Research Academy

Equations for computation of production, and demand

$$H1_{prod} = max\left(a^{H1}_{prod} * H1_{prod\ correction} - b^{H1}_{prod} * wages - c^{H1}_{prod} * [H1_{def} + H1_i - \bar{H1}], 0\right)$$

$$P1_{prod} = max\left(a^{P1}_{prod} * P1_{prod\ correction} - b^{P1}_{prod} * wages - c^{P1}_{prod} * [P1^{total}_{def} + P1_i - \bar{P1}], 0\right)$$

$$P1HH_{reg} = num_{reg} * demand_{P1HH} * correction_{P1demand}$$
$$H1HH_{reg} = num_{reg} * demand_{H1HH} * correction_{H1demand}$$
$$P1HH_{aff} = num_{aff} * demand_{P1HH} * correction_{P1demand} * P1_{affluence\_factor}$$
$$H1HH_{aff} = num_{aff} * demand_{H1HH} * correction_{H1demand} * H1_{affluence\_factor}$$

20

FROM LINEAR TO CIRCULAR INTEGRATED WASTE MANAGEMENT SYSTEMS: A FRAMEWORK

Authors: Selene Cobo, Antonio Dominguez-Ramos, Angel Irabien

*cobos@unican.es, +34 942 20 09 31

04/10/2017

Sosprocan DePRO Desarrollo de Procesos Químicos y Control de Contaminantes

WCCE10 ECCE11 ECAB4 1st - 5th October, 2017 Barcelona, Spain

Departamento de Ingenierías Química y Biomolecular

UC UNIVERSIDAD DE CANTABRIA

---

1. Motivation

Resource consumption

1.64 planets

(Global footprint network, 2016)

2

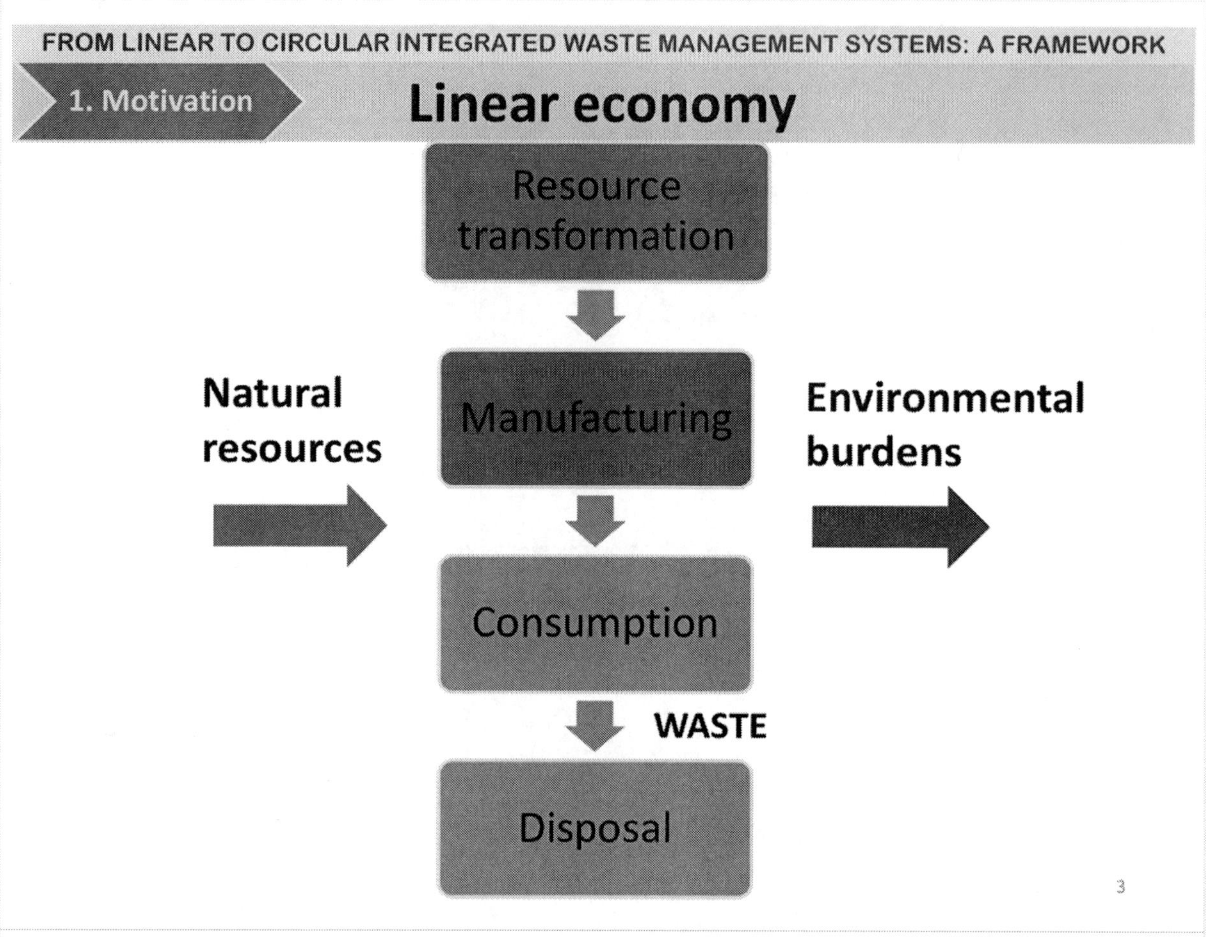
FROM LINEAR TO CIRCULAR INTEGRATED WASTE MANAGEMENT SYSTEMS: A FRAMEWORK
1. Motivation
Linear economy
Resource transformation
Natural resources
Manufacturing
Environmental burdens
Consumption
WASTE
Disposal
3

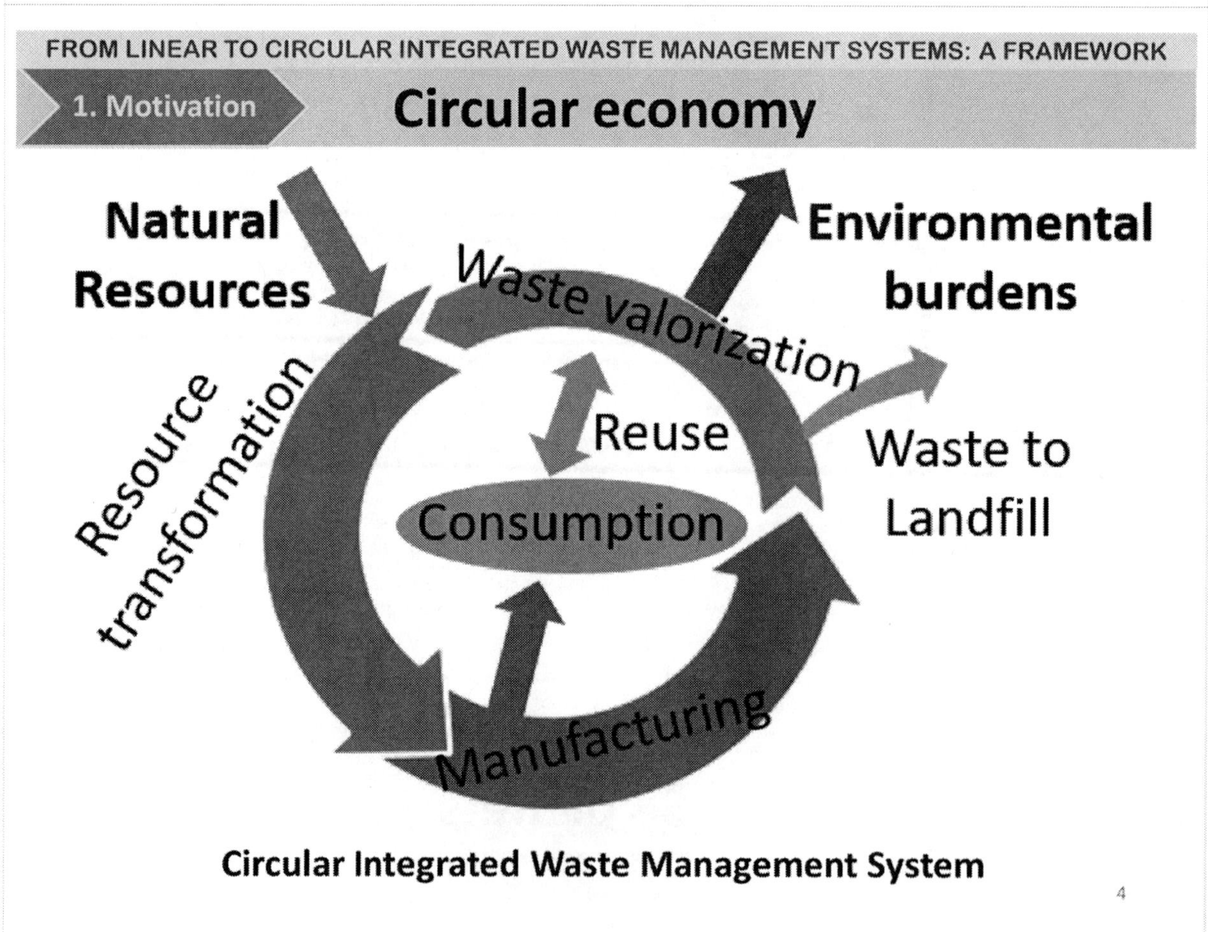
FROM LINEAR TO CIRCULAR INTEGRATED WASTE MANAGEMENT SYSTEMS: A FRAMEWORK
1. Motivation
Circular economy
Natural Resources
Environmental burdens
Waste valorization
Resource transformation
Reuse
Waste to Landfill
Consumption
Manufacturing
Circular Integrated Waste Management System
4

Aim of the work

To propose a methodological approach that allows decision-makers to identify the configuration of a Circular Integrated Waste Management System (CIWMS) with the minimal environmental impacts and consumption of resources.

2. Background

Limitations of recycling

- Pollutant accumulation
- Disassembly technologies
- Long distances to reach final destination

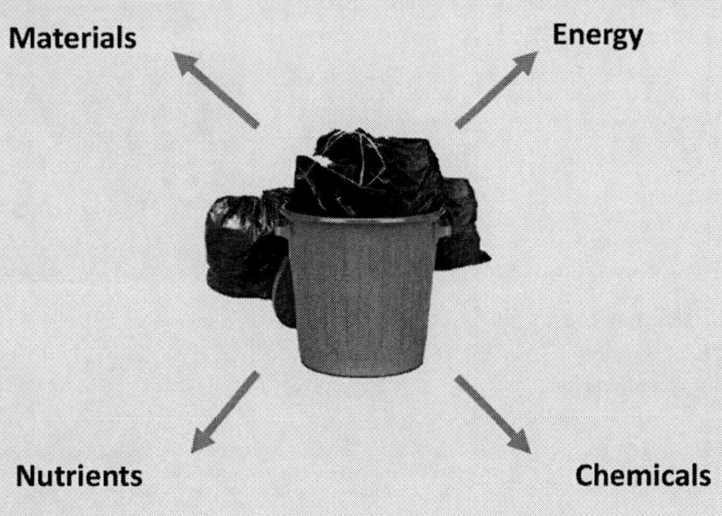

FROM LINEAR TO CIRCULAR INTEGRATED WASTE MANAGEMENT SYSTEMS: A FRAMEWORK

3. Definition

Linear system

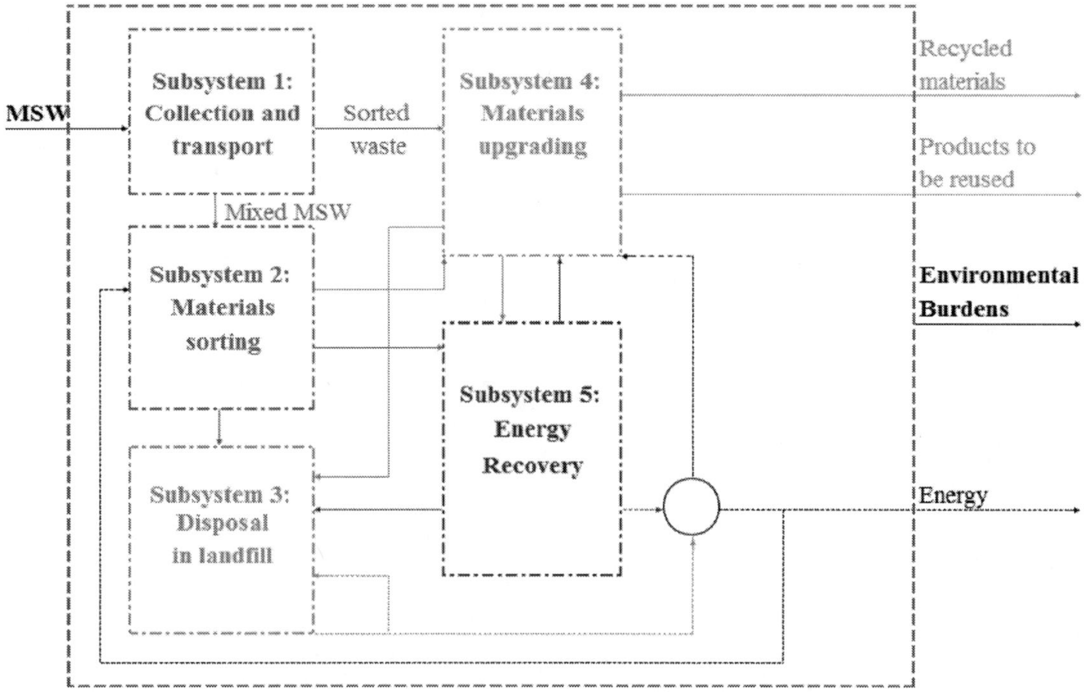

Linear Integrated Waste Management System

7

FROM LINEAR TO CIRCULAR INTEGRATED WASTE MANAGEMENT SYSTEMS: A FRAMEWORK

3. Definition

Upstream processes

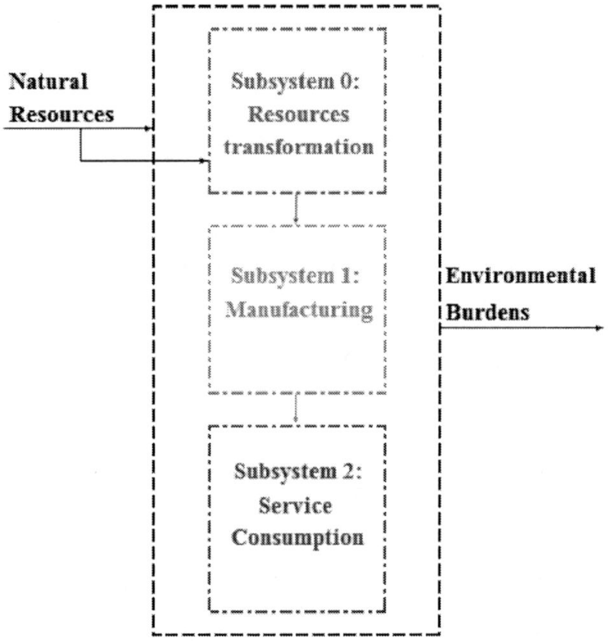

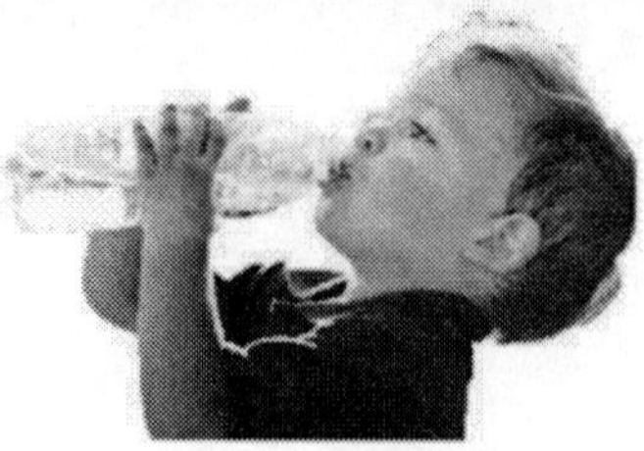

8

FROM LINEAR TO CIRCULAR INTEGRATED WASTE MANAGEMENT SYSTEMS: A FRAMEWORK

3. Definition

Circular system

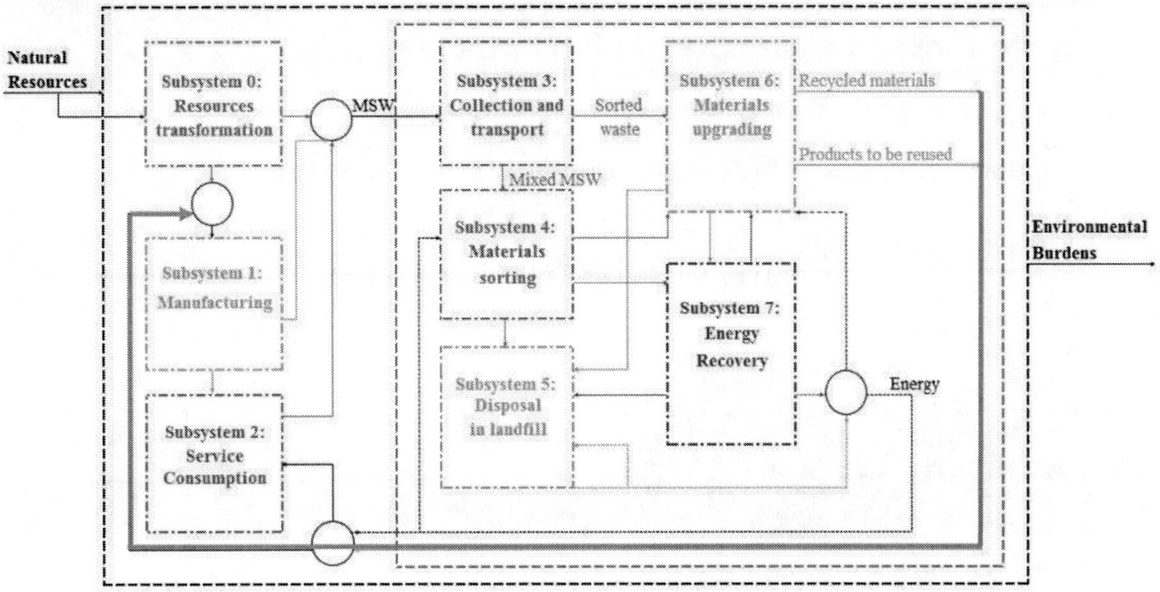

Cobo et al. 2017.*Resour. Conserv. Recycl.*

------ Linear IWMS

------ CIWMS

Circular Integrated Waste Management System

9

FROM LINEAR TO CIRCULAR INTEGRATED WASTE MANAGEMENT SYSTEMS: A FRAMEWORK

4. Tools

Methodological tools

Material Flow Analysis

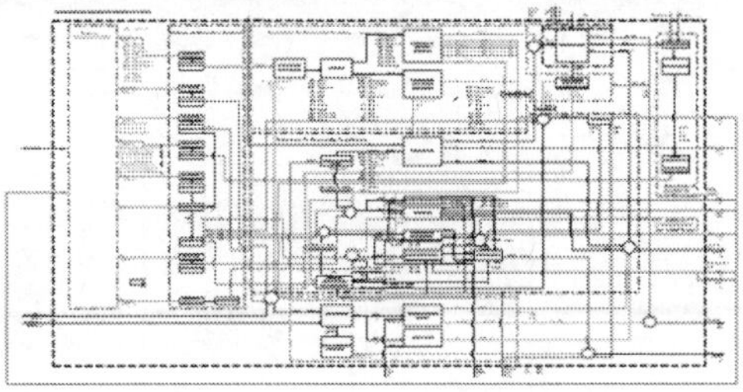

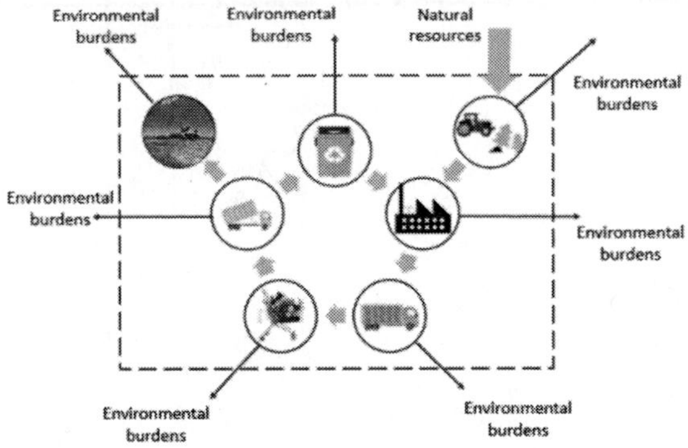

Life Cycle Assessment

Optimization

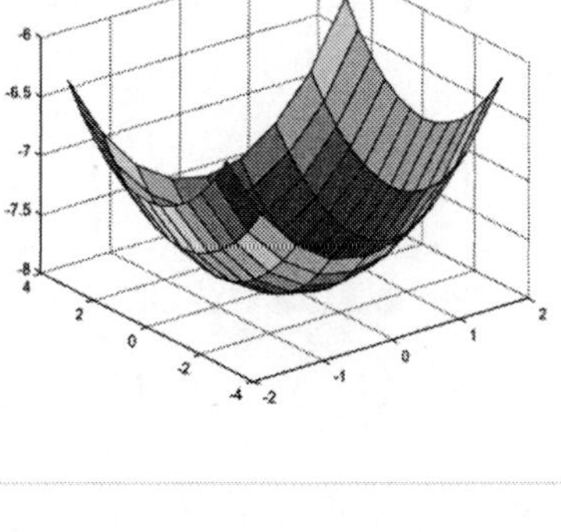

10

5. Advantages

Waste prevention

How to consider waste prevention in the analysis?

i) System boundaries that include upstream processes

ii) Functional unit that reflects the prevented amount of waste

From waste to resource

11

FROM LINEAR TO CIRCULAR INTEGRATED WASTE MANAGEMENT SYSTEMS: A FRAMEWORK

5. Advantages

Quantification of biogenic CO_2

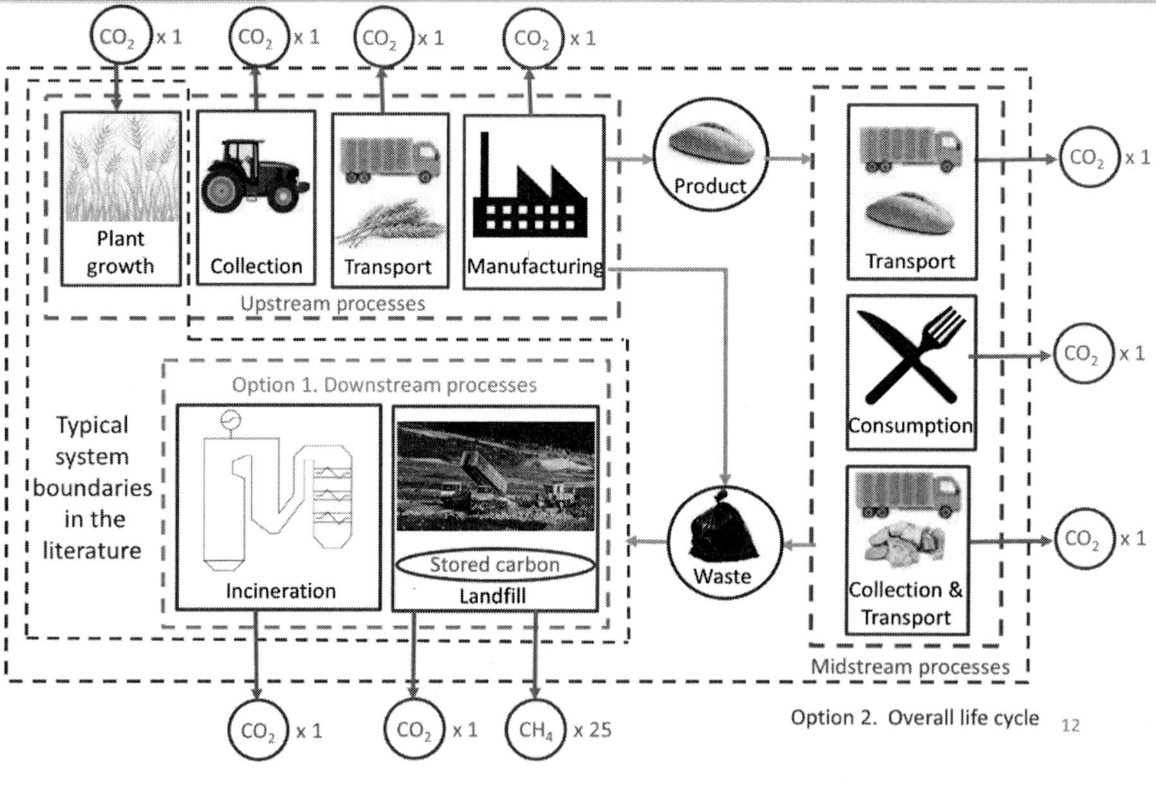

12

6.

Conclusions

- Expansion of system boundaries
- Multiple possible configurations→ design + analysis tools
- Benefits of the CIWMS cannot be taken for granted.

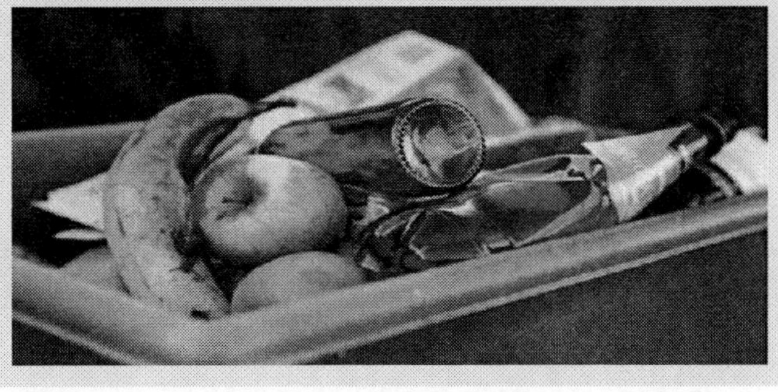

13

6.

Future work

Case study: valorization of the OW in Cantabria

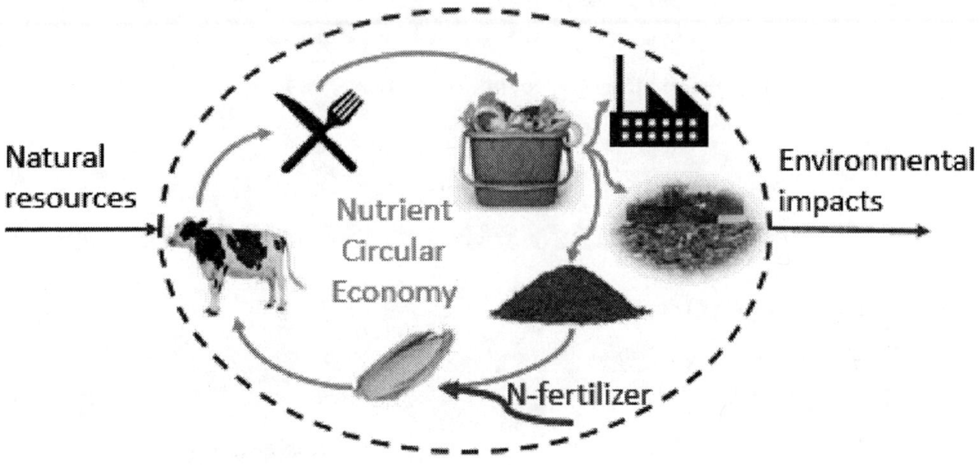

14

FROM LINEAR TO CIRCULAR INTEGRATED WASTE MANAGEMENT SYSTEMS: A FRAMEWORK

Acknowledgements

FPU research contract, code FPU15/01771

GOBIERNO DE ESPAÑA | MINISTERIO DE EDUCACIÓN, CULTURA Y DEPORTE

Thank you for your attention!!

Sosprocan DePRO Desarrollo de Procesos Químicos y Control de Contaminantes

WCCE10 ECCE11 ECAB4 1st - 5th October, 2017 Barcelona, Spain

FROM LINEAR TO CIRCULAR INTEGRATED WASTE MANAGEMENT SYSTEMS: A FRAMEWORK

Authors: Selene Cobo, Antonio Dominguez-Ramos, Angel Irabien

***cobos@unican.es, +34 942 20 09 31**

04/10/2017

Sosprocan DePRO Desarrollo de Procesos Químicos y Control de Contaminantes

WCCE10 ECCE11 ECAB4 1st - 5th October, 2017 Barcelona, Spain

Departamento de Ingenierías Química y Biomolecular

UC UNIVERSIDAD DE CANTABRIA

AUTHOR INDEX